Communications in Computer and Information Science 1692

More information about this series at https://link.springer.com/bookseries/7899

Xiaomin Ying (Ed.)

Human Brain and Artificial Intelligence

Third International Workshop, HBAI 2022, Held in Conjunction with IJCAI-ECAI 2022, Vienna, Austria, July 23, 2022
Revised Selected Papers

Editor
Xiaomin Ying
Beijing Institute of Basic Medical Sciences
Beijing, China

ISSN 1865-0929 ISSN 1865-0937 (electronic)
Communications in Computer and Information Science
ISBN 978-981-19-8221-7 ISBN 978-981-19-8222-4 (eBook)
https://doi.org/10.1007/978-981-19-8222-4

This Springer imprint is published by the registered company Springer Nature Singapore Pte Ltd.
The registered company address is: 152 Beach Road, #21-01/04 Gateway East, Singapore 189721, Singapore

Preface

The International Workshop on Human Brain and Artificial Intelligence (HBAI) endeavors to present and discuss original research at the intersection of brain and cognitive science, neural computation, artificial intelligence (AI), brain-computer interface, and data science along with its applications. This year HBAI was held in conjunction with the 31st International Joint Conference on Artificial Intelligence and the 23rd European Conference on Artificial Intelligence (IJCAI-ECAI 2022).

We received more than 20 full paper submissions, most of which provided interesting and valuable new insights related to human brain and artificial intelligence. Each submission was single-blindly reviewed by at least three reviewers. After peer review, HBAI 2022 featured 19 contributions that were carefully selected from all submissions. The different topics of the accepted papers were presented and discussed at a one-day seminar. To aid the discussion of those interesting insights the seminar was divided into four sessions covering topics such as AI for brain-related data analysis, AI and brain interfaces, and brain-related research.

HBAI 2022 represented a diligent effort from researchers in the area of the human brain and AI. We are very grateful to the 104 authors that submitted their papers and to the 23 members of the Program Committee (PC), who invited reviewers, produced reviews, and participated in the decision-making process. The PC members, who are listed on the following page, were absolutely critical to the workshop's success.

In addition to those who contributed to the content of HBAI 2022, we also want to thank the people dedicated to providing services for our one-day seminar. Guohua Dong, Yaowen Chen, Shuofeng Hu, Zhen He, Jinhui Shi, Runyan Liu, and Sijing An made significant efforts to support this workshop and they deserve a big thanks!

Organizing HBAI 2022 was interesting and challenging work for the committee, and we greatly hope that all participants had an unforgettable time as we did.

September 2022 Xiaomin Ying

Organization

General Chair

Xiaomin Ying	Beijing Institute of Basic Medical Sciences, China

Organizing Committee

Xiaomin Ying	Beijing Institute of Basic Medical Sciences, China
Yiwen Wang	Hong Kong University of Science and Technology, China
An Zeng	Guangdong University of Technology, China
Yuanyuan Mi	Chongqing University, China

Program Committee

Amy Wenxuan Ding	Emlyon Business School, France
Boqing Gong	Google Inc., USA
Daqing Guo	University of Electronic Science and Technology of China, China
Zengcai Guo	Tsinghua University, China
Yong Hu	University of Hong Kong, China
Kun Huang	Indiana University School of Medicine, USA
Wanzeng Kong	Hangzhou Dianzi University, China
Shurong Liu	Global Biotech Inc, USA
Yuanyuan Mi	Chongqing University, China
Dan Pan	Guangzhou Benzhen Networks Technology Co. Ltd., China
Hui Shen	National University of Defense Technology, China
Yiyu Shi	University of Norte Dame, USA
Aniruddha Sinha	Tata Consultancy Services, India
Dong Song	University of Southern California, USA
Xiaowei Song	Simon Fraser University, Canada
Huajin Tang	Zhejiang University, China
Jimin Wang	Google Inc., USA
Yalin Wang	Arizona State University, USA
Yiwen Wang	Hong Kong University of Science and Technology, China
Minpeng Xu	Tianjin University, China

Xiaomin Ying — Beijing Institute of Basic Medical Sciences, China
An Zeng — Guangdong University of Technology, China
Daoqiang Zhang — Nanjing University of Aeronautics and Astronautics, China

Contents

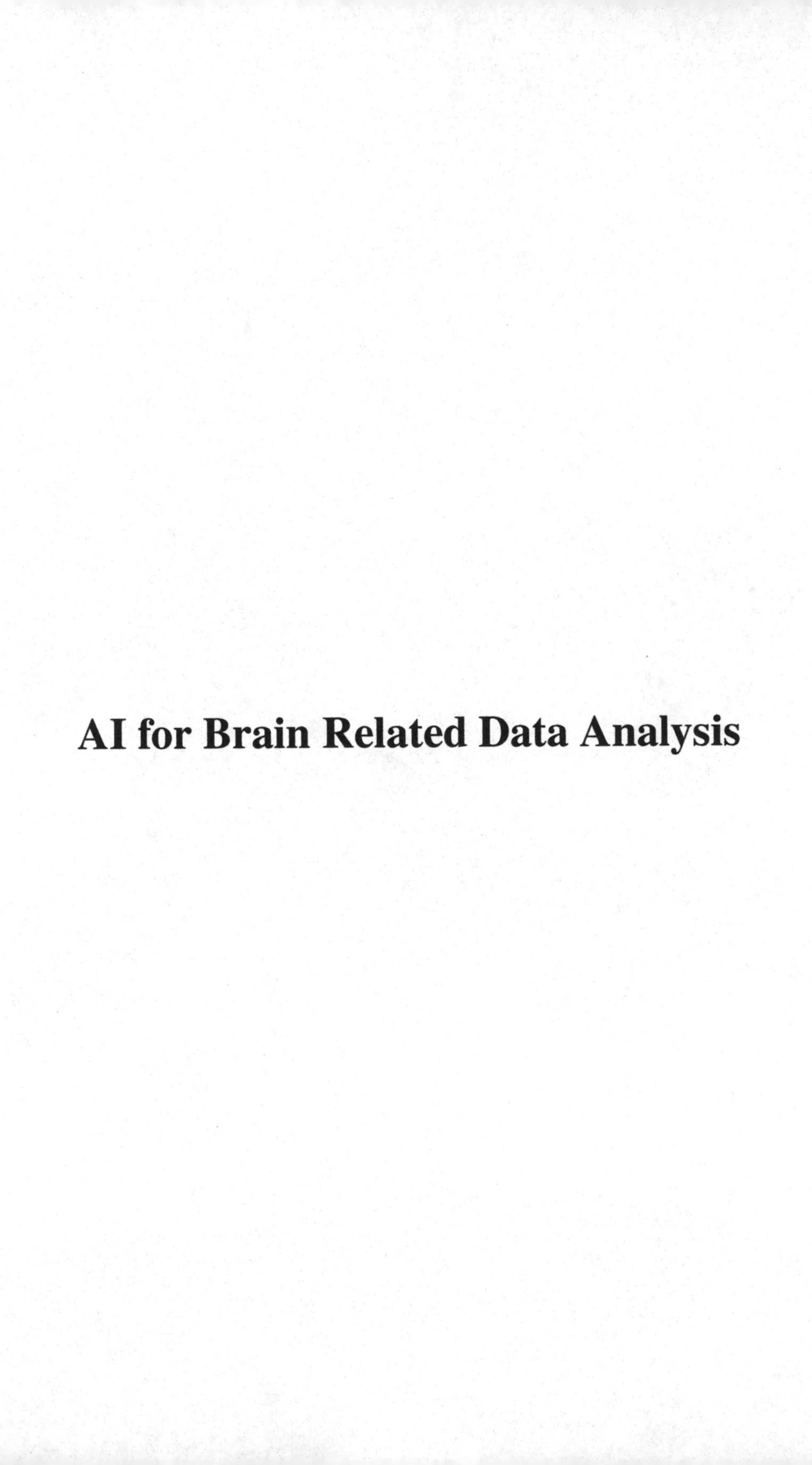

AI for Brain Related Data Analysis

Classification of EEG Signals Based on GA-ELM Optimization Algorithm

Weiguo Zhang[1], Lin Lu[2(✉)], Abdelkader Nasreddine Belkacem[3], Jiaxin Zhang[1], Penghai Li[1], Jun Liang[4], Changming Wang[5], and Chao Chen[1,6]

[1] Tianjin University of Technology, Tianjin 300384, China
[2] Zhonghuan Information College Tianjin University of Technology, Tianjin 300380, China
lulin1020@outlook.com
[3] Department of Computer and Network Engineering, College of Information Technology, UAE University, Al Ain 15551, UAE
belkacem@uaeu.ac.ae
[4] Tianjin Medical University General Hospital, Tiajin 300052, China
Evanliangjun@tmu.edu.cn
[5] Beijing Key Laboratory of Mental Disorders, Beijing Anding Hospital, Capital Medical University, Beijing 100088, China
[6] Academy of Medical Engineering and Translational Medicine, Tianjin University, Tianjin 300072, China

Abstract. There are many unpredictable problems in motion visualization and observation in BCI system, such as interference from external noise and visual fatigue of subjects. These problems seriously affect the performance of the whole BCI system. To solve this problem, this paper designed the experimental paradigm of imagination and observation, and built the eeg acquisition platform by combining UNITY and MATLAB. Ten healthy subjects participated in the experiment, which was divided into two stages: in the first stage, each subject was required to perform five experiments at the same time. In the second stage, after an interval of more than one month, the eeg signals of the 10 subjects were collected again (the same experimental paradigm). In pattern recognition and Hilbert huang transform time and frequency domain characteristics of extreme learning machine recognition classification based on genetic algorithm, and using the basic method of SVM algorithm and ELM comparison between the results and draw HHT and optimization algorithm of single collection of experiment acquisition signal has a significant effect, high classification rate can reach 85.3%.

Keywords: BCI · Motor imagination · Motor observation · Hilbert huang transform · ELM

1 Introduction

Motor imagery (MI) refers to the behavior of imagining a specific action without actually performing the action, which has been widely concerned in neuroscience and other fields (Hetu et al. 2013). MI activity in the brain generates various forms of signals that can

X. Ying (Ed.): HBAI 2022, CCIS 1692, pp. 3–14, 2023.
https://doi.org/10.1007/978-981-19-8222-4_1

be measured in different ways. Electroencephalogram (EEG) measures scalp electrical activity generated by the brain, providing non-invasive, high temporal resolution, and low cost solutions (Edgar et al. 2020). The brain-computer interface (BCI), based on motor imagery EEG (MI-EEG) signals, allows users to establish a direct channel between their brain and external devices. Complete the tasks of controlling the dual-arm robot and controlling the UAV to drive the virtual car (Liu et al. 2019, Nourmohammadi et al. 2018).

Classification algorithm is the key to distinguish limb MI, which is directly related to the accuracy of classification results, and is of great significance in various applications of MI. However, there are some inherent characteristics (such as non-stationarity) that are not conducive to classification of MI-EEG signals, which bring some challenges to the classification algorithm.

At present, the commonly used classifiers include self-organizing Map (SOM), support vector machine (SVM), k-nearest Neighbor (KNN), etc. Through a series of research and analysis, the classification accuracy has been significantly improved, but there is still some content to be improved in mi-EEG signal classification algorithm. For example, the problem of how to get the classification results quickly while maintaining the classification accuracy is helpful to realize the online MI-EEG signal classification with high accuracy and promote the real implementation of BCI system.

SVM is a variety of common kernel functions including linear kernel function and polynomial kernel function, gaussian kernel function, the choice of kernel function for the performance of its performance has a crucial role, by controlling the kernel function can determine decision boundary of linear or nonlinear, linear kernel function of support vector machine has the advantage of low complexity, fast speed,The advantage of nonlinear kernel support vector machine is that it can better fit the boundary between different classes in many cases. Support vector machine (SVM) is a fast and reliable classification algorithm, which can complete the classification task well under the condition of limited amount of data, so it is widely used in the research of eeg signal recognition (Lin et al. 2010).

Huang et al. proposed a novel machine learning algorithm of single hidden layer feedforward neural network, namely extreme Learning Machine (ELM). At the beginning of the training of the learning machine, the bias and input weights of its hidden layer are randomly given, and no adjustment is required in the training process. Compared with back Propagation (BP) neural network, extreme learning machine not only avoids falling into local optimal solution, but also has the advantages of strong generalization ability and fast training speed, so it has a good application prospect. Zhang L, Wen D et al. achieved an accuracy of 84.38% in the classification process of extreme learning machine (Zhang et al. 2020), but it took a long time in the classification process.

According to the above mentioned problem of how to improve classification efficiency while ensuring accuracy, this paper uses Genetic Algorithm (GA) to optimize extreme learning machine for classification processing of eeg signals.

2 Optimization of Extreme Learning Machine by Genetic Algorithm

Typical single hidden layer feedforward neural network is composed of input layer, hidden layer and output layer, and the input layer and hidden layer, hidden layer and output layer neurons are fully connected.

There are m neurons in the input layer, corresponding to M input variables; The hidden layer has N neurons; The output layer has H neurons corresponding to H output variables.

Let the connection weight matrix ω between the input layer and the hidden layer be:

$$\omega = \begin{bmatrix} \omega_{11} & \cdots & \omega_{1m} \\ \vdots & \ddots & \vdots \\ \omega_{1n} & \cdots & \omega_{nm} \end{bmatrix}_{n\times m} \tag{1}$$

wherein, ω_{ji} represents the connection weight between the i_{th} neuron in the input layer and the J_{th} neuron in the hidden layer.

Suppose that the connection weight β between the hidden layer and the output layer can be expressed as:

$$\beta = \begin{bmatrix} \beta_{11} & \cdots & \beta_{1h} \\ \vdots & \ddots & \vdots \\ \beta_{n1} & \cdots & \beta_{nh} \end{bmatrix}_{n\times h} \tag{2}$$

β_{jk} represents the connection weight between the JTH neuron in the hidden layer and the KTH neuron in the output layer. Assume that the threshold b of neuron of hidden layer is:

$$b = \begin{bmatrix} b_1 \\ \vdots \\ b_n \end{bmatrix}_{n\times 1} \tag{3}$$

Let the activation function of hidden layer neurons be g(x), then the output T of the network is:

$$T = \begin{bmatrix} t_1 & \cdots & t_m \end{bmatrix}_{1\times m} \tag{4}$$

Can be expressed as:

$$H\beta = T^T \tag{5}$$

where, T^T is the transpose of the matrix T; H is the hidden layer output matrix of the neural network. After input weights and hidden layer thresholds are determined randomly, the hidden layer output matrix is uniquely determined. Thus the output weight matrix can be obtained:

$$\overline{\beta} = H^+T \tag{6}$$

where H^+ is the generalized inverse of the output matrix H of the hidden layer.

Genetic Algorithm (GA) is an efficient global optimization search Algorithm based on natural selection and Genetic theory, which combines the survival rule of the fittest in the process of biological evolution with the random information exchange mechanism of chromosomes in the population. The problem parameters are encoded as chromosomes, and the selection is carried out in an iterative way. Crossover and mutation operations exchange chromosome information in the population, so that the population evolves to better and better regions in the search space from generation to generation, until the optimal solution point is reached.

The traditional ELM algorithm also has some shortcomings, especially poor performance in the face of high-dimensional noise data, because the single-layer structure of ELM is not enough for feature learning. Deep learning can obtain data features of different levels through continuous learning of original data features, which can reduce the dimension of data and remove the interference and noise in the data, so as to extract high-level features of the original data. Therefore, the extension of ordinary single-layer ELM to multi-layer structure can better learn the essential characteristics of data. On the basis of this idea, in order to improve the classification accuracy of ELM, this paper proposed a GA-based optimization ELM (GA-ELM). In this algorithm, input layer weight and hidden layer node threshold of ELM are mapped to genes on each chromosome in GA population, and each chromosome is encoded. The chromosome fitness of GA corresponds to the classification accuracy of ELM. The optimal chromosome was selected through GA. When the chromosome evolved to the preset maximum number of iterations, the optimal chromosome in the population was selected as the input weight and hidden layer threshold of the optimized ELM, so as to classify the data set. The algorithm effectively integrates the parallel searching ability of GA and the super fitting ability of

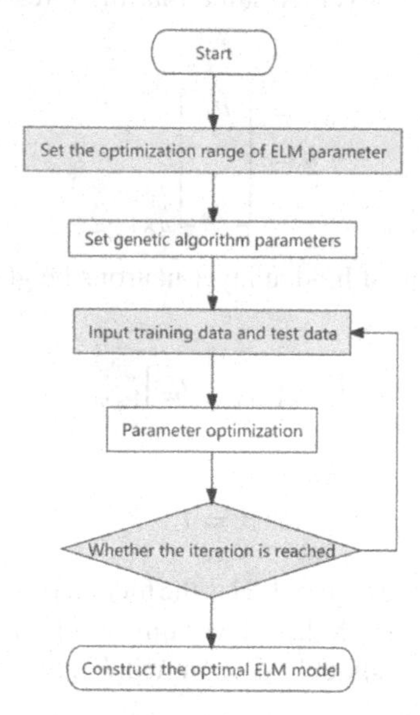

Fig. 1. Optimization process of ELM parameters by genetic algorithm

ELM. Figure 1 shows the optimization process of extreme learning machine by genetic algorithm.

3 The Experiment Design

In this experiment, each participant needs to carry out two stages of the experiment. In the first stage, each participant carries out 5 groups of experiments in total, and the experimental scene of each group is repeated 12 times, including the exercise imagination and observation. Each subject completed a total of 5 × 24 scenes in one experiment. In stage 2 (at an interval of more than one month), all the steps of Stage 1 are repeated. A total of 10 healthy people aged between 20 and 25 were selected for this experiment. All the subjects learned and understood relevant safety knowledge and experimental nature before the experiment.

3.1 Experimental System Framework

The framework of this experimental system is as follows: the MATLAB software running program of the main computer (EEG acquisition) sends out the signal to control the playback of virtual reality scene, and the unity3d software of another auxiliary computer (virtual scene simulation) receives and processes the signal. The video of arm movement observation and arm sneaker imagination simulation (lever holding) will be played in the unity scene, The subject can also freely control the movement of the small ball (for later debugging tasks). While controlling the small ball, the subject needs to observe and imagine the movement of the single hand grip according to the prompt signal. The control flow chart of the experiment is shown in Fig. 2.

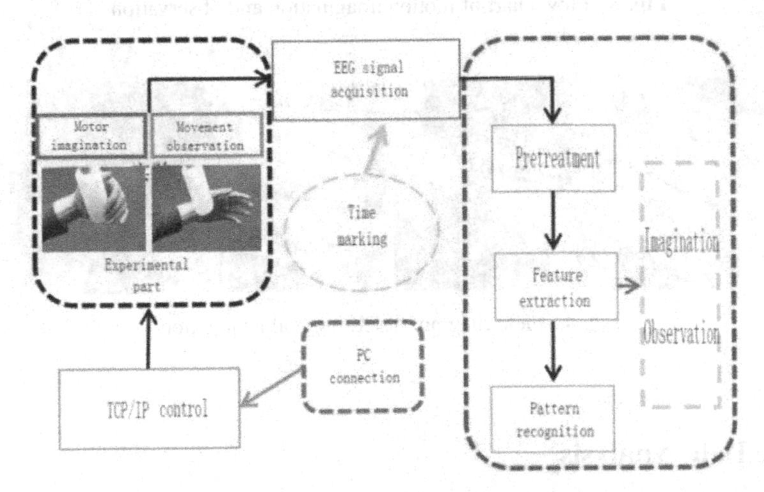

Fig. 2. Experimental control flow chart

3.2 Data Acquisition

The flow of data acquisition experiment is shown in Fig. 3: an experiment includes three stages: prompt stage, experimental stage and rest stage. The experimental stage includes motion imagination stage and motion observation stage, and the rest stage is maintained for 2S. According to the motor observation cue (2 s), the subjects performed the brain observation activities of seeing the rod and holding the hand; In the motor imagination stage, the subjects carried out the brain motor imagination activity of holding the rod by arm according to the motor imagination prompt information (2 s). The duration of each experiment was set at 20.620 s. In addition, the experiment also had the autonomous operation link of the left and right hands to control the movement of the ball. Each subject was collected for 5 times in total (5 times can be separated if the condition is not good). Each experiment included 12 motion observation and 12 motion imagination stages, and the total length of the experiment was 6 min and 59 s. There will be rest and prompt interface in the experiment, and the subjects need to perform different tasks according to the different prompt information (Fig. 4).

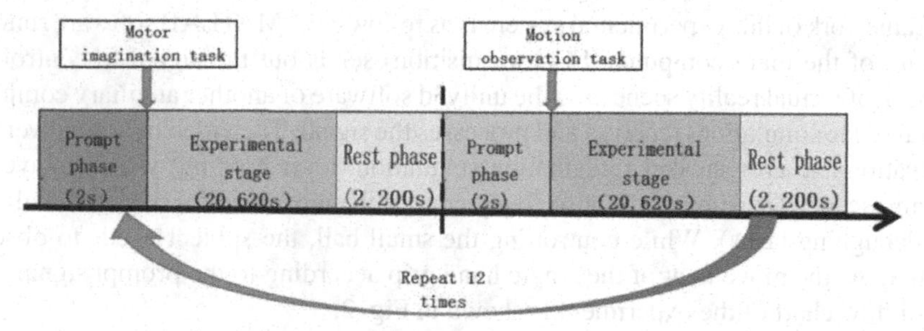

Fig. 3. Flow chart of motion imagination and observation

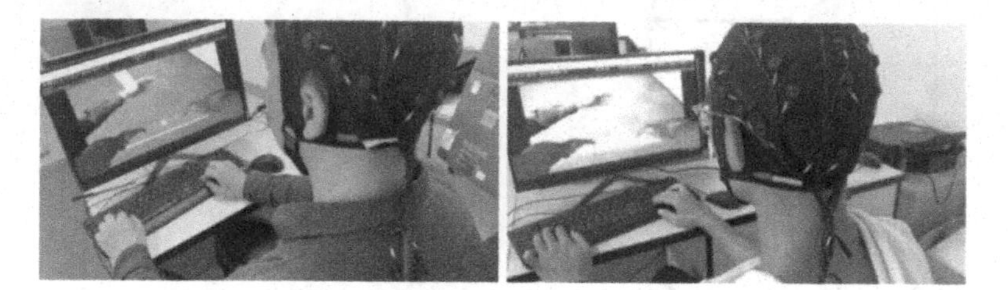

Fig. 4. Field diagram of EEG signal acquisition

4 The Data Analysis

4.1 Preprocessing

The eeg signals required in this experiment are mainly concentrated between 0.5 and 30 Hz. Before the analysis of EEG signals, we need to filter the data and remove the interference of power frequency signals and other noises. In terms of filter selection, compared with other filters, butterworth filter has stable amplitude-frequency characteristics both in and out of the passband. In view of stability, this experiment uses a 0.530 Hz fourth-order Butterworth filter for band-pass filtering.

4.2 Feature Extraction

Empirical mode decomposition (EMD) can adaptively decompose the original EEG signal into many IMF components. According to the different IMF components, the characteristics contained in the original signal are also different. A subject needs five experiments, while one experiment needs 12 normal motor imagination and 12 motor observation. The original signal x (T) after EMD can be expressed as:

$$x(t) = \sum_{i=1}^{n} w_i(t) + r_n(t) \tag{7}$$

where n represents the number of times to filter IMF components, I represents the number of times, CI (T) represents the i-th IMF component, and RN (T) represents the remaining component.

After each IMF component of the signal x (T) obtained by the above EMD decomposition, the IMF component Si (T) of each order can be subjected to Hibert transformation. The transformation function is as follows:

$$\mathrm{Y}_i(t) = \frac{1}{\pi} \int_{-\infty}^{+\infty} \frac{\mathrm{C}_i(\tau)}{t - \tau} \tag{8}$$

According to the Hilbert spectrum obtained by HHT transformation of the first three order IMF, the marginal spectrum can be obtained by continuing Hilbert transformation. The figure below shows the marginal spectrum of motion imagination and motion observation after the superposition of the average of the same events on a selected subject's C3 and C4 channels (Fig. 5).

The above figure shows the marginal spectrum of AO and Mi events of C3 and C4 channels of the same subject respectively, and the marginal spectrum of C3 channel of the subject collected again one month later. By comparing the marginal spectra of two states (AO and MI) in the same period, we can see that on C3 channel, after the first red line (10 Hz), the energy in MI and AO states decreases significantly, and ERD phenomenon occurs. The amplitude of MI near the red line and yellow line is less than that of Ao. On C4 channel, the energy of AO and MI at 815 Hz near the red line and 1530 Hz near the yellow line shows an upward trend, that is α Ers phenomenon occurs in all bands. One month later, the blue line and yellow line on the C3 channel of the subject also showed the ERD phenomenon of energy decline, but the amplitude fluctuation was not as large as that one month ago, which proved that the experimental state of the subject one month later was not as good as that of the subject one month ago.

In this paper, the marginal spectrum of each event is windowed. The window length and window shift are 1 Hz. 30 eigenvectors are obtained for each event, so the subjects have 24 events, so the eigenvectors are 24 * 5 * 30 in total.

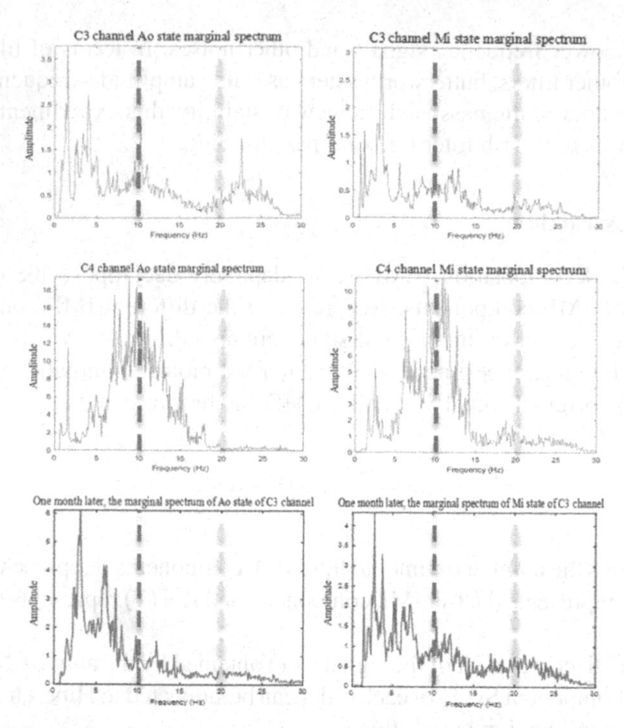

Fig. 5. Marginal spectrum of the same subject under different events and times

4.3 Genetic Algorithm Optimized Parameter Setting

According to the above optimization process, the parameters of the genetic algorithm are selected, the crossover probability Px and mutation probability Pm are 0.7 and 0.01 respectively, the maximum evolution algebra MAXGEN is set as 100, the number of individuals is set as 20, and the sample number PN of the training set is 70% of the total SN of the data set. TN, the sample number of the test set, is 30% of the total sample SN of the data set. Further, by optimizing and recording the optimal value of each generation, the weight matrix and paranoia were finally obtained, and the two parameters were put into the ELM network for training.

5 Results

In this experiment, a total of 4 s eeg signals were extracted. The sampling frequency of the signal is 1024 Hz, so each event contains $4*1024 = 4096$ original data points. Hilbert transformation was used to obtain $30*24*5 = 3600$ feature points (one channel). The motion imagination and motion observation experiments selected C3 and C4 channels, so each method had 1440 and 7200 features respectively. HHT characteristics of 10 subjects were input into KNN, SVM, ELM and GA-ELM classification respectively.Finally, the classification accuracy of GA-ELM classifier is higher than the other three classifiers, with an accuracy of 85.3%.

6 Discussion

According to the mean and standard deviation of classification accuracy in Table 1 of the first stage, it can be seen that the classification effect of GA-ELM optimization algorithm is significantly higher than that of the other three classifiers. Although the results of test 5 and 8 in GA-ELM were poor, the overall classification accuracy was 10% higher than KNN classifier, 5% higher than SVM classifier, and 4% higher than ELM classifier. The classification accuracy of test 1 was the highest, up to 82.5%. Based on GA-ELM optimization algorithm, the classification effect of subjects 1, 2, 3, 4, 6, 7, 9 and 10 was better, and the lowest classification rate of 10 subjects reached more than 78.3%. The classification rate of subjects 5 and 8 is not ideal, because each subject has some differences more or less in the process of collecting EEG, such as the speed of brain receiving information and the degree of focus.

Table 1. Classification accuracy of three methods. Classification accuracy of MI and AO (AVG ± S.D.)

Subject number	KNN	SVM	ELM	GA-ELM
1	0.732	0.783	0.792	0.825
2	0.695	0.762	0.746	0.813
3	0.687	0.785	0.766	0.798
4	0.625	0.692	0.789	0.801
5	0.681	0.753	0.724	0.783
6	0.733	0.783	0.746	0.808
7	0.682	0.662	0.721	0.792
8	0.736	0.765	0.732	0.785
9	0.703	0.697	0.713	0.812
10	0.729	0.773	0.794	0.803
Average	0.701 ± 0.032	0.750 ± 0.042	0.752 ± 0.029	0.800 ± 0.013

In the second stage, the eeg signals of 7 out of 10 subjects were re-collected at an interval of more than one month (same experiment). Feature extraction and classification go through the same steps as above, and the classification accuracy is shown in Table 2.

According to the mean and standard deviation of classification accuracy in Table 2 of the second stage, it can be seen that the classification effect of GA-ELM optimization algorithm is still significantly higher than that of the other three classifiers. The overall classification accuracy is 11% higher than KNN, 6% higher than SVM and 5% higher than ELM. The classification accuracy of test 1 was the highest, reaching 85.3%. Based on GA-ELM optimization algorithm, the classification effect of subjects 1, 2, 3 and 4 was better, and the lowest classification rate of 7 subjects reached more than 78.5%.

According to the comparison between the classification accuracy of seven subjects in Table 2 and Fig. 6 in the second stage, the classification accuracy of the subjects

Table 2. Classification accuracy of three methods after one month. Classification accuracy of MI and AO (AVG ± S.D.)

Subject number	KNN	SVM	ELM	GA-ELM
1	0.737	0.783	0.781	0.853
2	0.673	0.692	0.754	0.825
3	0.705	0.766	0.734	0.800
4	0.676	0.663	0.721	0.815
5	0.673	0.733	0.793	0.798
6	0.718	0.795	0.753	0.785
7	0.667	0.775	0.786	0.799
Average	0.692 ± 0.026	0.750 ± 0.043	0.760 ± 0.025	0.810 ± 0.022

one month after test 1, 2, 3, 4, 5 and 7 is higher than that of the previous experiment, which indicates that the subjects are in better experimental state and the training is more effective after one month. The accuracy of tested 6 decreased, but the overall average increased by 1%. The lowest classification accuracy of GA-ELM reached 78.5%, which was 0.2% higher than the lowest one month ago. The highest classification accuracy of tested 1 reached 85.3%. This shows that after a month of familiarity with the experiment, the effect of eeg acquisition is better. The experimental data of two stages verified the validity of GA-ELM in classification of motor imagination and observed EEG signals, and also proved the separability of motor imagination and observation.

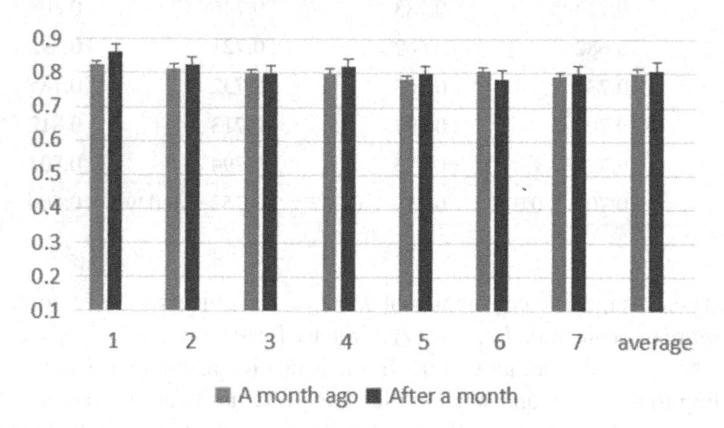

Fig. 6. Comparison of experimental classification accuracy before and after one month

7 Conclusion

In this paper, we study the movement of the imagination and observation of brain-computer interface system, by taking ten subjects of eeg signals, preprocessing and feature extraction, and then the characteristic value input to the limit of the machine learning classifier are optimized by the genetic algorithms, will eventually get the classification results were compared with those of KNN, SVM and traditional ELM classifier, the final classification result was 85.3%. The experimental results show that the proposed method is feasible. Although this experiment make up some shortages of the traditional ELM algorithm, but there are still some deficiencies, due to the limitation of economic conditions, this experiment is mainly collected health of young people, has failed to live up to the analysis of different population of the age, more so for different data sets its own characteristics and ELM complex relationship between number of hidden layer neurons, still needs further research.

References

Hetu, S., Gregoire, M., Saimpont, A., et al.: The neural network of motor imag ery: an ALE meta-analysis. Neuroscience Biobehavioralre Views **37**(5), 930–949 (2013)

Edgar, P., Torres, E.A., Hernándex-Lvarez, M., et al.: EEG-based BCI emotion recognition: a survey. Sensors **20**(18), 5083 (2020)

Liu, Y., Su, W., Li, Z., et al.: Motor-imagery-based teleoperation of a dual-arm robot performing manipulation tasks. IEEE Trans. Cognitive Dev. Syst. **11**(3), 414–424 (2019)

Nourmohammadi, A., Jafari, M., Zander, T.O.: A survey on unmanned aerial vehicle remote control using brain–computer interface. IEEE Trans. Human-Machine Syst. **48**(4), 337–348 (2018)

Zhang, L., Wen, D., Li, C., et al.: Ensemble classifier based on optimized extreme learning machine for motor imagery classification. J. Neural Eng. **17**(2), 026004.1–026004.12 (2020)

Huang, G.B., Zhu, Q.Y., Siew, C.K.: Extreme learning machine: theory andapplications. Neurocomputing **70**(1–3), 489–501 (2006)

Adams, I.L.J., Lust, J.M., Steenbergen, B.: Development of motor imagery ability in children with developmental coordination disorder–a goal-directed pointing task. Br. J. Psychol. **109**(2), 187–203 (2018)

Hortal, E., Planelles, D., Costa, A., et al.: SVM-based Brain-Machine Interface for controlling a robot arm through four mental tasks. Neurocomputing **151**(1), 116–121 (2015)

Pfurtscheller, G.: EEG rhythms-event-related desynchronization and synchronization. Rhythms in Physiological Systems. Springer, Berlin, Heidelberg, pp. 289–296 (1991). https://doi.org/10.1007/978-3-642-76877-4_20

Roy, R., Konar, A., Tibarewala, D.N., et al.: EEG driven model predictive position control of an artificial limb using neural net. In: 2012 Third International Conference on Computing, Communication and Networking Technologies (ICCCNT' 12). IEEE, pp. 1–9 (2012)

Berg, L.P., Vance, J.M.: An Industry case study:investigating early design decision making in virtual reality. J. Computing Information Science in Eng. (6), 1001–1006 (2016)

Cortes, C., Vapnik, V.: Support vector machines. Mach. Learn. **20**, 273–293 (1995)

Vapnik, V.N., Chervonenkis, A.: A note on one class of perceptrons. Automation and Remote Control **25**(1), 821837 (1964)

Khemchandani, R., Chandra, S.: Twin support vector machines for pattern classification. IEEE Trans. Pattern Anal. Mach. Intell. **29**(5), 905–910 (2007)

Li, Z., Du, M.: HHT based lung sound crackle detection and classification. In: Proceedings of 2005 International Symposium on Intelligent Signal Processing and Commimication Systems,Hong Kong, 11, pp. 13–16 (2005)

Khosrowabadi, R., Quek, H.C., Wahab, A., et al.: EEG-based emotion recognition using self-organizing map for boundary detection. In: 2010 20th International Conference on Pattern Recognition, Istanbul, Aug 23-26, 2010. Piscataway: IEEE, pp. 4242-4245 (2010)
Lin, Y.P., Wang, C.H., Wu, T.L., et al.: EEG-based emotion recognition in music listening. IEEE Transaction on Biomedical Engineering, **57**(7), 17981806 (2010)

Delving into Temporal-Spectral Connections in Spike-LFP Decoding by Transformer Networks

Huaqin Sun[1], Yu Qi[2,3], and Yueming Wang[1(✉)]

[1] Qiushi Academy for Advanced Studies, Zhejiang University, Hangzhou, China
{sunhuaqin,ymingwang}@zju.edu.cn

[2] Affiliated Mental Health Center and Hangzhou Seventh People's Hospital, Zhejiang University School of Medicine, Hangzhou, China
qiyu@zju.edu.cn

[3] MOE Frontier Science Center for Brain Science and Brain-Machine Integration, Zhejiang University School of Medicine, Hangzhou, China

Abstract. Invasive brain-computer interfaces (iBCIs) have demonstrated great potential in neural function restoration by decoding intention from brain signals for external device control. Spike trains and local field potentials (LFPs) are two typical intracortical neural signals with good complementarity from time and frequency domains. However, existing studies mostly focused on a single type of signal, and the interaction between the two signals has not been well studied. This study proposes a temporal-spectral transformer network (TSNet) to model the temporal (with spikes), spectral (with LFPs), and mutual (with both signals) connections in spike-LFPs towards robust neural decoding. Experiments with clinical neural signals demonstrate that the attention-based connection model enables the dynamic temporal-spectral compensation in spike and LFP signals, which improves the robustness against temporal shifts and noises in neural decoding.

Keywords: Brain-computer interfaces · Spike-LFP fusion · Neural decoding

1 Introduction

Invasive brain-computer interface (iBCI) enables direct communication with external devices from neural activities, which has demonstrated great potential in clinical applications such as motor function restoration and neuroprostheses [4,5,9,14,20].

One crucial problem of BCI is how to extract effective motor information from neural signals. From the perspective of invasive BCIs, typical neural signals include action potentials (spikes) and local field potentials (LFPs). The two types of signals encode motor-related information differently. With spike signals, motor information is encoded by the timing and frequency of a single neuron's

X. Ying (Ed.): HBAI 2022, CCIS 1692, pp. 15–29, 2023.
https://doi.org/10.1007/978-981-19-8222-4_2

firing activities [7,8,11,17]. With LFP signals, which reflect the activities of a large number of neurons, the spectral domain contains rich information for motor decoding [16]. The two signals demonstrate strong complementarity. Spike signals contain accurate temporal information, while they are usually sensitive to noises and suffer from low stability. On the contrary, LFP signals are more stable in time while demonstrating lower resolution both temporally and spatially [10]. Therefore, the fusion of both sides can potentially improve both the accuracy and robustness of neural decoding [3].

Several studies have explored the fusion of spikes and LFPs to leverage the information from both signals. [3] extracted spike counts from spikes and spectral features from LFPs and selected and combined the features from both signals for neural decoding. [1] utilized a point process model to extract features from spike signals and a Gaussian process model to extract LFP features and merge both features in neural decoding. [18] fused the features of spikes and LFPs by feature concatenation and applied the broad learning system for intention decoding. These studies demonstrated the effectiveness and necessity of the fusion of spikes and LFPs, while they mostly concatenated the two signals directly and ignored the interaction between them. Especially, the temporal-spectral connections in spike and LFP signals can dynamically change given different tasks and conditions [15,23]. Existing approaches used a fixed combination between spikes and LFPs, leading to suboptimal performance.

This study investigates the temporal-spectral interaction between spike and LFP signals with a transformer neural network. It first learns the temporal and spectral connections within the spike and LFP signals, respectively, and then captures the temporal-spectral mutual information between spike and LFP signals. With the attention mechanism in transformer networks, our model can dynamically adjust the inter- and intra- connections in spikes and LFPs to cope with changes in signals. Experiments with clinical neural signals show that modeling the interaction between spike and LFP signals helps improve the robustness of neural decoding against temporal shifts and noises, and achieves the state-of-the-art performance.

2 Methods

We propose a temporal-spectral transformer network (TSNet) to model the dynamic temporal-spectral connections in spike and LFP signals, as illustrated in Fig. 1. The TSNet contains four main modules: 1) a temporal self-attention component to learn the intra-connections within spike trains; 2) a spectral self-attention component to describe the intra-connections in LFP signals; 3) a temporal-spectral cross-attention component to model the interaction between the spike and LFP signals; and 4) a classification layer to learn task-related representations. As shown in Fig. 1, raw neural signals are firstly filtered and processed to spike and LFP signals, respectively. Then spike and LFP signals go through the temporal self-attention component and the spectral self-attention component, respectively, and then interact in the cross-attention component. Finally, task-related representations are extracted for neural decoding tasks.

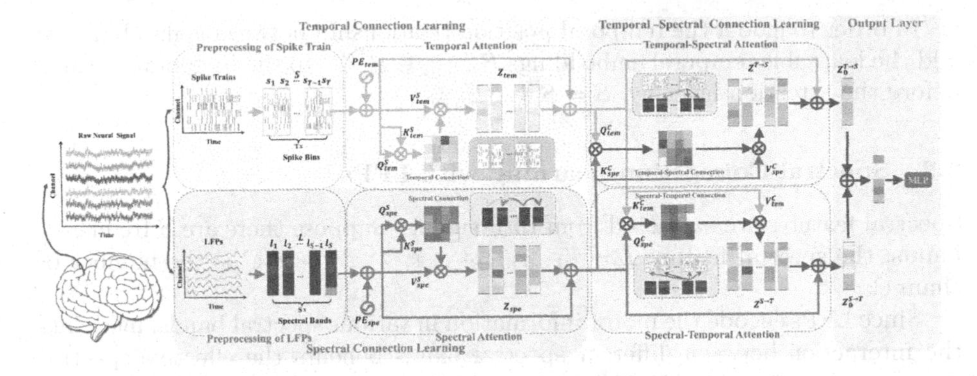

Fig. 1. The framework of the temporal-spectral transformer network (TSNet).

2.1 Temporal Connection Learning with Spikes

We extract temporal features of spikes by computing the spike counts of non-overlap bins to obtain a feature sequence of $S = [s_1; \ldots; s_T] \in \mathbb{R}^{T \times C}$, where T denotes the number of bins in a sample and C represents the number of channels on the signal recorder. $s_t \in \mathbb{R}^{1 \times C}$ denotes the spike count of C channels in the bin t.

Since a motor behavior usually involves a sequence of movements, the interaction between temporally adjacent spike bins contains rich information for accurate motor decoding. We model the temporal connections between spike bins using the self-attention mechanism [21] in transformer networks to capture the dynamic relationship patterns among spike bins. With the transformer network framework, we should learn three representations of query, key, and value, which can be computed by:

$$Q_{tem}^S = SW_Q^S, K_{tem}^S = SW_K^S, V_{tem}^S = SW_V^S, \tag{1}$$

where $W_Q^S \in \mathbb{R}^{C \times d_q}$, $W_K^S \in \mathbb{R}^{C \times d_k}$ and $W_V^S \in \mathbb{R}^{C \times d_v}$ are the corresponding linear transformation matrice. Then the temporal connections from spike bin s_t to all spike bins $[s_1; \ldots; s_T]$ can be defined by the similarity of corresponding query q_t and keys K_{tem}^S computed by scaled dot-product. After that, we model the interaction among the spike bins based on the self-attention mechanism. Specifically, we define the attention feature z_t from spike bin s_t to all spike bins S by a weighted sum of the values V_{tem}^S, where the weights are the temporal connections. Thus, we obtain the temporal attention among spikes bins by:

$$\begin{aligned} Z_{tem} &= Attention_{tem}(Q_{tem}^S, K_{tem}^S, V_{tem}^S) \\ &= \text{softmax}\left(\frac{Q_{tem}^S K_{tem}^S}{\sqrt{d_q}}\right) V_{tem}^S, \end{aligned} \tag{2}$$

where $Z_{tem} = [z_1; \ldots; z_T] \in \mathbb{R}^{T \times d_v}$ denotes the temporal attention feature learned by the self-attention layer based on spikes. Finally, a fully connected feed-forward layer with a residual connection and layer normalization is applied.

In order to model the temporal position relationship between spikes bins, we add the learnable temporal embeddings $PE_{tem} \in R^{T \times C}$ to the temporal features before the attention layer by: $S = S + PE_{tem}$.

2.2 Spectral Connection Learning with LFPs

Spectral features are extracted from LFP signals. Suppose there are S frequency bands, the spectral feature $L = [l_1; \dots ; l_S] \in \mathbb{R}^{S \times C}$, where C is the number of channels.

Since LFPs encode the motor information in various spectral bands, modeling the interaction between different spectral bands benefits the effective spectral representation learning. Similar to the modeling of temporal connections, we learn the interaction between spectral bands using the self-attention network by:

$$\begin{aligned} Z_{spe} &= Attention_{spe}(Q^S_{spe}, K^S_{spe}, V^S_{spe}) \\ &= \text{softmax}\left(\frac{Q^S_{spe} K^S_{spe}}{\sqrt{d_q}}\right) V^S_{spe}, \end{aligned} \tag{3}$$

where $Z_{spe} = [z_1; \dots ; z_S] \in \mathbb{R}^{S \times d_v}$ denotes the spectral attention feature learned by the self-attention layer based on LFPs.

The position information between different spectral bands $PE_{spe} \in R^{S \times C}$ is learned by: $L = L + PE_{spe}$.

2.3 Temporal-Spectral Connection Learning with Spike-LFPs

The temporal and spectral representations are then fused with a cross-attention model [13] to learn temporal-spectral connections in spike-LFP signals.

Specifically, we first learn the linear transformations from temporal Z_{tem} and spectral representations Z_{spe} to queries (Q^C_{tem}, Q^C_{spe}), keys (K^C_{tem}, K^C_{spe}) and values (V^C_{tem}, V^C_{spe}) by:

$$\begin{aligned} Q^C_{tem} = Z_{tem} W_Q^{C_{tem}}, K^C_{tem} = Z_{tem} W_K^{C_{tem}}, V^C_{tem} = Z_{tem} W_V^{C_{tem}}, \\ Q^C_{spe} = Z_{spe} W_Q^{C_{spe}}, K^C_{spe} = Z_{spe} W_K^{C_{spe}}, V^C_{spe} = Z_{spe} W_V^{C_{spe}}. \end{aligned} \tag{4}$$

After that, we model the connections from the temporal domain to the spectral domain with the similarity of queries from spikes (Q^C_{tem}) and key-value pairs from LFPs ($K^C_{spe} - V^C_{spe}$). The cross-attention mechanism attends to the spectral information conditioned on the temporal feature, which models the connection from the temporal to the spectral domain. We obtain the temporal-spectral attention feature by $Attention_{T \to S}$:

$$\begin{aligned} Z^{T \to S} &= Attention_{T \to S}(Q^C_{tem}, K^C_{spe}, V^C_{spe}) \\ &= \text{softmax}\left(\frac{Q^C_{tem} K^C_{spe}}{\sqrt{d_q}}\right) V^C_{spe}. \end{aligned} \tag{5}$$

We also model the connection from the spectral to the temporal domain utilizing a similar way by $Attention_{S \to T}$:

$$\begin{aligned} Z^{S \to T} &= Attention_{S \to T}(Q^C_{spe}, K^C_{tem}, V^C_{tem}) \\ &= \mathrm{softmax}\left(\frac{Q^C_{spe} K^C_{tem}}{\sqrt{d_q}}\right) V^C_{tem}. \end{aligned} \tag{6}$$

The schematic diagram of $Attention_{T \to S}$ and $Attention_{S \to T}$ lies in the top and bottom, respectively, in the subfigure of Temporal-Spectral Connection Learning component in Fig. 2.

2.4 Task-Related Output Layer

In order to fit the task, we prepare two learnable classification embeddings $S_{cls} \in \mathbb{R}^{1 \times C}$, $L_{cls} \in \mathbb{R}^{1 \times C}$ added in front of the respective features of each signal, and get the corresponding final output $z_0^{T \to S} \in \mathbb{R}^{1 \times d_S}$, $z_0^{S \to T} \in \mathbb{R}^{1 \times d_L}$ after TSNet as the representations, following the way of BERT [6]. Thus the two embeddings contain rich intra-modal and inter-model information, and we get the final fusion representation by adding for simplicity. A multilayer perceptron (MLP) is applied to classify different tasks from the sum of $z_0^{T \to S}$, $z_0^{S \to T}$ by:

$$\hat{Y} = f_{MLP}(z_0^{T \to S} + z_0^{S \to T}), \tag{7}$$

$$\mathcal{L} = \ell\left(Y, \hat{Y}\right), \tag{8}$$

where ℓ is the cross-entropy loss function, and $\hat{Y}$ is the prediction from MLP.

3 Experiments and Results

Here we evaluate and analyze the temporal-spectral connections learned by the TSNet using a clinical dataset with a human subject.

3.1 Clinical Dataset

Neural Signal Acquisition. A paralyzed participant was implanted with two 96-channel Utah intracortical microelectrode arrays (Blackrock Microsystems, Salt Lake City, UT, USA) in the left primary motor cortex to record the neural signals. Raw neural signals were sampled at 30 kHz using the Neuroport system (NSP, Blackrock Microsystems) with a 96-channel Utah intracortical microelectrode array. All clinical and experimental procedures in this study were approved by the Medical Ethics Committee of The Second Affiliated Hospital of Zhejiang University (Ethical review number 2019-158, approved on 05/22/2019).

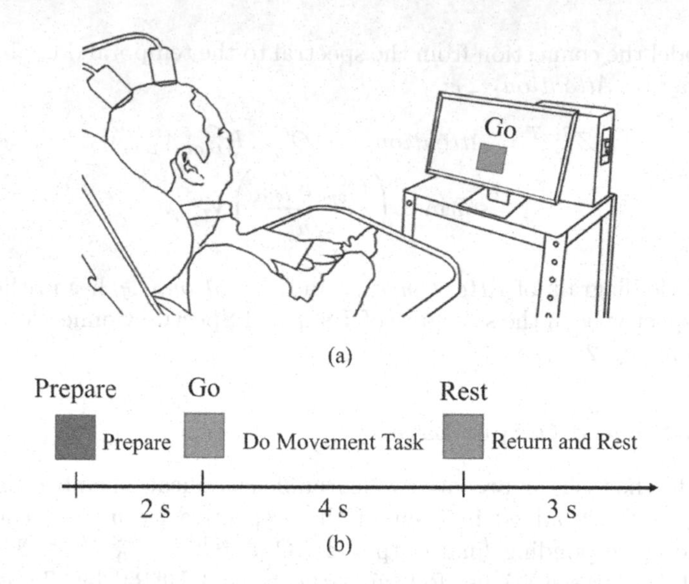

Fig. 2. The experiment paradigm. (a) Neural data collection with a human participant. (b) The timeline of a single trial.

Experimental Paradigm. The experimental paradigm includes performing ten movements: EyebrowsRaise, MouthOpen, HeadTurnUp, HeadTurnRight, RightArmRaise, RightElbowFlex, RightWristExt, RightHandOpen, RightToeTip, RightKneeLift. In each trial, the participant was asked to watch the computer monitor and repeat the movements represented in short films following the movement instruction (see Fig. 2(a)). Specifically, the participant sequentially prepared for motor commands, imagined or performed the commanded movements, and rested (see Fig. 2(b)). The participants can perform the first four movements physically and imagine the last six movements. Each movement was repeated ten times each day, and the clinical dataset included neural signals of five experimental days.

Neural Signal Processing. We extract temporal and spectral features from spikes and LFPs, respectively. For spike signals, raw neural signals were filtered by a high-pass filter (250 Hz cut-off). Then spikes were detected by a threshold of -6.5 to -4.5 times the root-mean-square value of the signals. A non-overlap sliding window with a length of 400 ms is used for temporal segmentation. Thus, a total of 10 bins are obtained from the "Go" period for each channel, and we get a temporal feature of $S \in \mathbb{R}^{10 \times 96}$ for each trial (with 96 channels). For LFP signals, we compute the spectral log-power in 10 consecutive 30 Hz bands from 0–300 Hz for each channel using the "Go" period similar to [19]. Therefore, for each trial, we obtain a spectral feature $L \in \mathbb{R}^{10 \times 96}$.

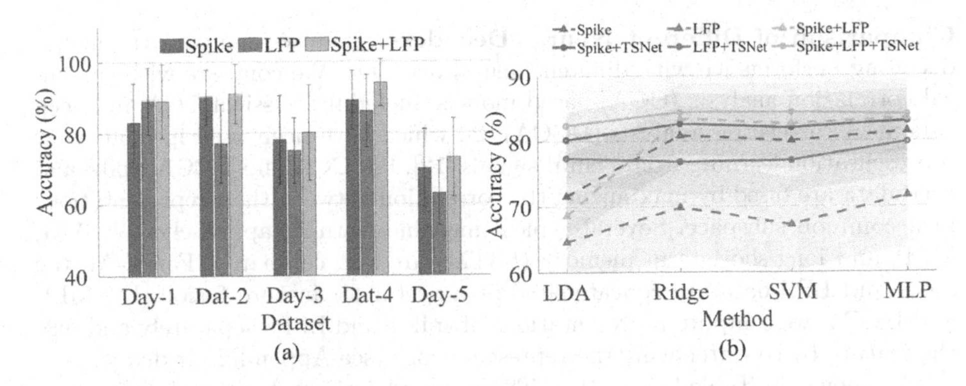

Fig. 3. Comparison of using spikes and LFPs alone and signal fusion. (a) Performance comparison with different experiment days. (b) Comparison of features with and without TSNet-based feature learning.

3.2 Spike-LFP Fusion Improves Neural Decoding Accuracy

Here we compare the neural decoding performance using spikes and LFPs, both individually and together. For TSNet, the hyperparameters of the dimension of query and value are selected by 10-fold cross-validations in $\{4, 8, 16\}$ and $\{31, 64, 128\}$, respectively. In model training, the batch size is 5, and the Adam algorithm is used with a learning rate of 10^{-3} and weight decay of 10^{-4}.

Effectiveness of Spikes-LFP Fusion. As shown in Fig. 3(a), for data from all five experimental days, the fusion of spike and LFP signals effectively improves the movement classification accuracy by up to 6% compared with using spike signals individually and by up to 14% compared with using LFPs individually. With the fusion of spikes and LFPs, the average classification accuracy is 85.40 %, which is significantly higher than using spikes (82.00%) and LFPs (78.00%), individually. We further evaluate the effectiveness of temporal-spectral connections by comparing the features with or without TSNet-based representation learning. In Fig. 3(b), solid lines represent features optimized with the TSNet, where Spike+TSNet is spike signals optimized using the temporal connections alone, and LFP+TSNet is LFP signals optimized using the spectral connections alone; and dash lines are the neural data without representation learning. A total of four classifiers are used, including linear discriminant analysis (LDA), ridge regression (Ridge), support vector machine (SVM), and multilayer perceptron (MLP). Overall, significant performance improvement is obtained with TSNet-based representation learning. For spikes, the performance improves by 1.6%–8.6% with different classifiers. For LFPs, the performance increases by 7.0%–12.8%. For the spike-LFP fusion, the performance improves by 0.6%–13.8%. The improvement is significant with the paired t-test. The result demonstrates the effectiveness of TSNet-based representation learning.

Comparison of Different Neural Decoders. Here we compare the neural decoding performance with different neural decoders. We compare with canonical correlation analysis (CCA)-based models, including classical CCA and deep canonical correlation analysis (DCCA) [2], which have been used in multi-view representation learning from neural signals [12]. For CCA and DCCA, spike and LFP data are fused by maximizing the correlation between their representations in a common subspace. Several typical machine learning approaches of SVM, MLP, and long short-term memory (LSTM) are also compared. For SVM, the spike and LFP data are concatenated as a vector for feature fusion; For MLP and LSTM, we compute representations of spikes and LFPs separately and fuse the feature by concatenating the representations (see Appendix for details).

As shown in Table 1, with the TSNet-based temporal-spectral connection learning, the overall accuracy is 85.4%, which is 2.9%, 2.4%, and 2.2% higher compared with SVM, MLP, and LSTM, respectively. Compared with CCA-based approaches, our method significantly outperforms CCA and DCCA by 5.6% and 4.4%, respectively. Results demonstrate that TSNet learns better fusion representations from spikes and LFPs.

Table 1. Comparison of different neural decoders.

	Spikes	LFPs	Spikes + LFPs
SVM	0.800 ± 0.043	0.672 ± 0.072	0.826 ± 0.061
MLP	0.814 ± 0.046	0.716 ± 0.079	0.834 ± 0.058
LSTM	$\mathbf{0.820 \pm 0.030}$	0.744 ± 0.050	0.836 ± 0.034
CCA [2]	-	-	0.798 ± 0.061
DCCA [2]	-	-	0.810 ± 0.067
TSNet(ours)	0.820 ± 0.042	$\mathbf{0.780 \pm 0.046}$	$\mathbf{0.854 \pm 0.072}$

3.3 Temporal Connections Improve Robustness to Temporal Shifts

The temporal-shift refers to the time difference of onset and duration in movement conduction among different trials, which has been a critical problem in accurate neural decoding. Temporal-shift is usually difficult to evaluate since the participant cannot perform most of the movement physically. Meanwhile, the existence of temporal-shift leads to difficulties in the accurate alignment of trials, thus degrading the neural decoding performance.

We find that the temporal connections can dynamically adapt to temporal-shift by emphasizing the temporal slots of movement conductions. Here we analyze how the self-attention adapts to temporal-shift, and evaluate the robustness of performance with temporal-shift conditions.

Attention Adaptation to Temporal Shifts. To analyze the problem of temporal shifts, we visualize the responses of neurons during a movement trial.

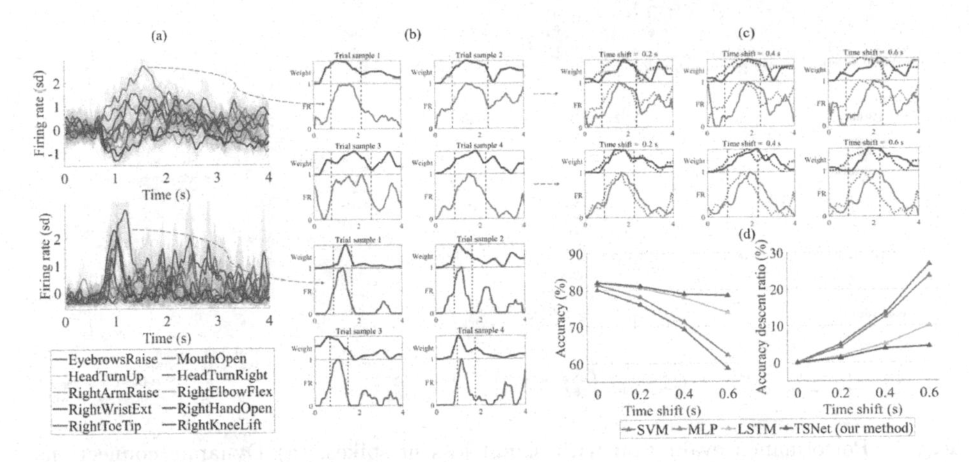

Fig. 4. Visualization of temporal attentions with temporal shifts. (a) Neural responses to different movements of two example neurons. The upper neuron responds strongly to "MouthOpen", and the lower neuron responds strongly to "RightToeTip". (b) Neural responses of example neurons in single trials and temporal connection weights correspondingly. The upper four subfigures are four trials of "MouthOpen", with orange lines indicating neural responses of the first example neuron and blue lines indicating temporal connection weights. The lower four subfigures are trials of "RightToeTip" and the neural responses of the second example neuron accordingly. (c) Comparison of neural responses and temporal attention weights with different temporal shifts (solid lines) and without temporal shifts (dotted lines). (d) Decoding performance with different conditions of temporal shifts.

Fig. 4(a) shows two neurons' responses to different movements. We find that the neurons respond strongly to certain movements after the "Go" cue. Specifically, the neuron in the upper subfigure of Fig. 4(a) shows a strong response to "MouthOpen" (the orange line), and the neuron in the lower subfigure responds strongly to "RightToeTip" (the purple line). The neurons' responses indicate the onsets of movements [24]. With the temporal-shift problem, the neural responses exhibit different onset time and durations.

Then we visualize the temporal attention weights predicted with the TSNet (upper subfigures) along with the responses of neurons (lower subfigures) in 4(b) (see Appendix for details). In each subfigure, the black dashed lines mark the start and end cues in a trial. We find that although the actual onset and duration of movements vary trial-by-trial, the temporal representations can dynamically focus on the movement durations, which benefits the robust representation learning against temporal shifts.

Performance with Temporal Shifts. Here we quantitively evaluate the robustness against temporal shifts. To simulate data with different temporal shift conditions and temporal variabilities, we manually control the temporal-shift condition by adjusting the alignment between neural signals and visual

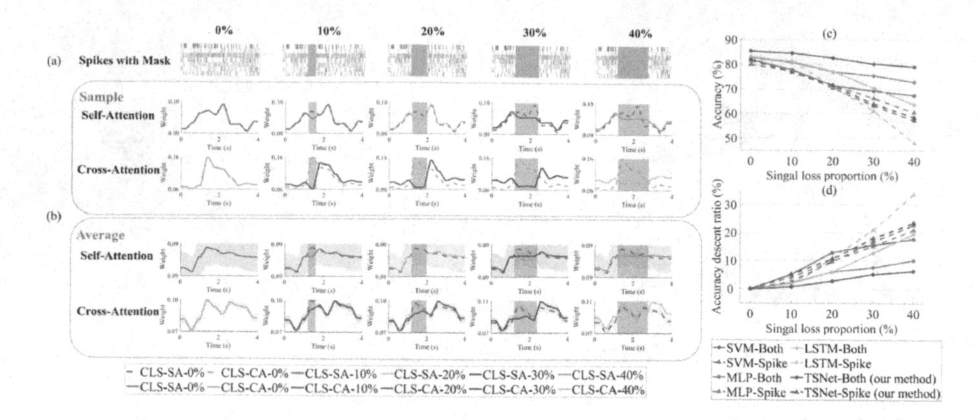

Fig. 5. Performance evaluation with signal loss in spikes. (a) Dynamic connections adapt to signal loss. The first row illustrates spike signals with zero masks. The second and third rows are the temporal (self-attention) and spectral-to-temporal (cross-attention) connection weights with different proportions of signal loss (solid lines) compared with weights without signal loss (dashed lines). (b) The average connection weights across trials. (c) Accuracy of different methods with signal loss. (d) Accuracy descent ratio of different methods with signal loss.

cues. Specifically, we add a 0.2 s, 0.4 s, or 0.6 s delay in the neural data of the test trials, then evaluate the neural decoding performance. Fig. 4(d) compares the neural decoding performance using different decoders. Overall, TSNet performs robustly against temporal shifts with only a 3.6% accuracy decrease even with a large shift of 0.6 s, where the performance of LSTM and MLP decreases by 6.56% and 19.2%, respectively. By visualizing the temporal attention weights with different temporal shifts in Fig. 4(c), we find that the TSNet can adapt to temporal shifts with the temporal connections, which improves the robustness against temporal shifts.

3.4 Temporal-Spectral Connections Improve Robustness to Noises

Spike and LFP signals contain respective properties in neural signal decodings. Spike signals are accurate but sensitive, and LFP signals can compensate for spikes to improve the robustness of decoders [3]. We find that the temporal-spectral connections learned by the TSNet can optimize the mutual representation of spike-LFP dynamically according to the effectiveness of signals. Here we simulate different conditions of signal loss to evaluate the dynamic compensation between spikes and LFPs with the temporal-spectral connections in TSNet.

Robustness Against Signal Loss. We simulate different conditions of signal loss by masking a proportion of neural spikes in the test trials. We simulate two types of signal loss: mask with zeros and mask with Gaussian noises. In Fig. 5(a), we illustrate the adaptation of the cross-attention in temporal-spectral representations with different conditions of signal loss (mask with zeros). In the subfigures of Fig. 5(a), the dashed lines indicate the temporal (self-attention) and spectral-to-temporal (cross-attention) connection weights without signal loss, and the solid lines are the weights given signal loss. The temporal connection weights drop at the slots containing signal loss and emphasize more on other slots as compensation. Meanwhile, the spectral-to-temporal connections decrease at the masked slots. Fig. 5(b) shows that the weight changes among different trials are mostly consistent. Similar results are obtained with the Gaussian noise (see Appendix). The results demonstrate that the TSNet improves the robustness of neural decoding by fusing spike and LFP signals with the dynamic temporal-spectral connections.

Then we evaluate the performance of TSNet with different conditions of signal loss in spikes. Results are shown in Fig. 5(c) and Fig. 5(d). Overall, the decoders using both spike and LFP signals (solid lines) demonstrate lower performance decreases with different proportions of signal loss, compared with using spikes alone (dashed lines). At the loss proportion of 40%, using fusion signals reduces the accuracy decrease by 8.2%, 10.8%, 14.2%, and 17.2% compared with using spikes alone, for SVM, MLP, LSTM, and TSNet, respectively. The TSNet achieves the highest accuracy and the lowest performance decrease with all the signal loss proportions. As shown in Fig. 5(c), TSNet achieves 79% of accuracy with the 40% loss proportion, outperforming LSTM and MLP by 15.4% and 6.2%, respectively. The results demonstrate the TSNet improves the robustness of neural decoding by fusing spike and LFP signals with the dynamic temporal-spectral connections.

4 Conclusion

This paper proposes a temporal-spectral transformer network (TSNet) to learn the dynamic temporal-spectral connections from spikes and LFPs. Clinical experimental results demonstrate that the attention network can dynamically adjust the connection weights to cope with changes in signals. The temporal connections can cope with time shifts in different trials to improve the accuracy and robustness of decoding. The temporal-spectral connections fuse the spike and LFP signals in a dynamic way that can improve the robustness to signal loss. These findings strongly suggest that the dynamic modeling of temporal-spectral connections is essential for accurate and robust neural decoding.

Acknowledgements. The authors would like to thank Kedi Xu, Junming Zhu and Jianmin Zhang for the support in the clinical experiments. This work was partly supported by the grants from National Key R&D Program of China (2018YFA0701400), Key R&D Program of Zhejiang (2022C03011), Natural Science Foundation of China (61906166, 61925603), the Fundamental Research Funds for the Central Universities, and the Starry Night Science Fund of Zhejiang University Shanghai Institute for Advanced Study (SN-ZJU-SIAS-002).

Appendix

A Detail Settings Of Neural Decoders

We compare the performance of different neural decoders, including SVM, MLP, LSTM, CCA, DCCA, and ridge regression. The detail model settings are as follows:

- **SVM:** We use one-vs-rest (ovr) for multiple class classification and the regularization parameter C is chosen from $\{10^{-6}, 10^{-4}, \cdots, 10^{6}\}$. Spikes and LFPs are flattened to vectors as model input. For the fusion of spikes and LFPs, we concatenate the two signals as vectors and feed them to SVM.
- **MLP:** We use a one-layer MLP and the hidden size is selected from $\{50, 100, 200, 300, 400, 500\}$. For fusion, we train two individual models on spikes and LFPs, respectively, and concatenate the representation from them as the input of a one-layer MLP with the hidden size selected form $\{50, 100, 200\}$.
- **LSTM:** We use one layer LSTM whose hidden size is selected from $\{50, 100, 200, 300, 400, 500\}$ to extract the sequence's representation and use the last hidden state as the final representation. The fusion strategy is the same as MLP.
- **CCA:** We select the linear projection dimension from $\{10, 20, 30, 40, 50, 100, 200\}$ and use an ECOC-SVM classifier as[22].
- **DCCA:** We use one layer neural network whose hidden size is selected from $\{50, 100, 200, 300, 400, 500, 800, 1000\}$ and the linear projection dimension and classification are the same as CCA.
- **Ridge regression:** The regularization strength α is chosen from $\{10^{-6}, 10^{-4}, \cdots, 10^{6}\}$.

Ten-fold cross-validation is used for parameter selection and performance evaluation. For the model training of MLP and LSTM, we set the batch size to 5 and use Adam as the optimizer with the learning rate of 10^{-3} and weight decay of 10^{-4}. The loss function is cross-entropy loss. Early stop is applied to avoid overfitting.

B Estimating Movement Conduction Durations With Neuron Responses

We use neurons' responses to estimate the true durations of movement intention conductions. We first select the neurons that respond strongly to specific motor tasks with the criterion of large response variation across the 'Go' period. Fig. 4(a) shows two example electrodes that tune to "MouthOpen" and "RightToeTip", respectively. Then we obtain the trial samples' data for the selected channel and task and do min-max normalization to transform the responses into decimals between 0 and 1 which showed in Fig. 4(b) with the corresponding color. We mark the time that the firing rate is higher than 0.5 as the start of motor intention. Then we compare the firing rate at 800ms after the start. If the firing rate is below 0.5, we set the end time; if not, we search for the time when the firing rate was below 0.5 as the end. We plot the black dashed lines as the start and end marks in a trial in Fig. 4 (b). The connection weights are plotted along with the firing rate using the blue line.

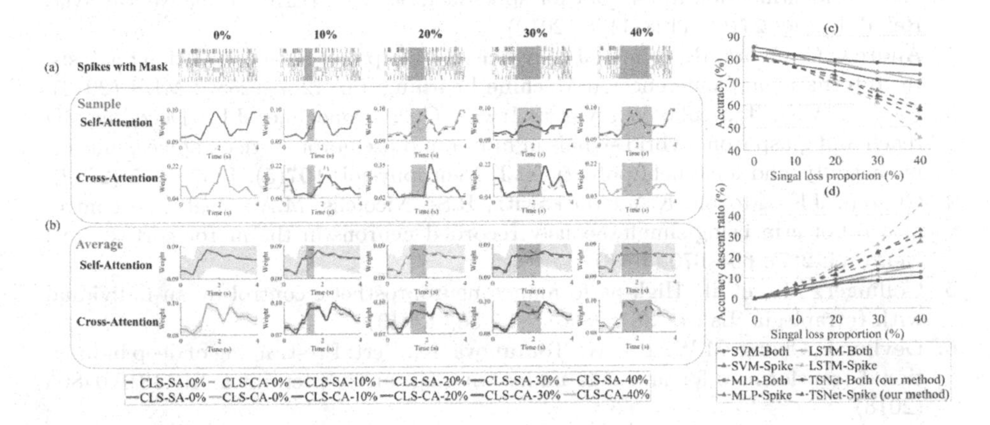

Fig. 6. Performance evaluation with Gaussian noises in spikes. (a) Dynamic connections adapt to noises. The first row illustrates spike signals with Gaussian masks. The second and third rows are the temporal (self-attention) and spectral-to-temporal (cross-attention) connection weights with different proportions of noises (solid lines) compared with weights without noises (dashed lines). (b) The average connection weights across trials. (c) Accuracy of different methods with Gaussian noises. (d) Accuracy descent ratio of different methods with Gaussian noises.

C Robustness To Gaussian Noises

Here, we analyze the result of masking a proportion of spike signals with Gaussian noise. In Fig. 6(a–b), we illustrate the adaptation of connection weights in temporal-spectral representations with different conditions of signal loss masking with Gaussian noise. Similar to results with signal loss using zero-masks, both the temporal (self-attention) and spectral-to-temporal (cross-attention) connection weights adjust dynamically to cope with changes in signals to improve the robustness and accuracy of neural decoding.

We evaluate the performance of the TSNet with different conditions of noises in spikes. Results are shown in Fig. 6(c) and Fig. 6(d). The results are consistent with the signal loss conditions. Firstly, the fusion of signals improves the robustness against noises. At the loss proportion of 40%, the accuracy descent using both signals outperform using spikes alone by 13.2%, 12.2%, 23.2%, and 16.2% in SVM, MLP, LSTM, and TSNet, respectively. Secondly, the TSNet achieves the best accuracy with high stability in spike-LFP fusion. As shown in Fig. 6 (c), TSNet achieves 77% of accuracy with the 40% loss proportion, outperforming

LSTM and MLP by 6.8% and 4.0%, respectively. TSNet also achieves the lowest accuracy descent ratio (Fig. 6(d)), demonstrating our method's robustness.

References

1. Abbaspourazad, H., Hsieh, H.L., Shanechi, M.M.: A multiscale dynamical modeling and identification framework for spike-field activity. IEEE Trans. Neural Syst. Rehabil. Eng. **27**(6), 1128–1138 (2019)
2. Andrew, G., Arora, R., Bilmes, J., Livescu, K.: Deep canonical correlation analysis. In: International Conference on Machine Learning, pp. 1247–1255. PMLR (2013)
3. Bansal, A.K., Truccolo, W., Vargas-Irwin, C.E., Donoghue, J.P.: Decoding 3D reach and grasp from hybrid signals in motor and premotor cortices: spikes, multiunit activity, and local field potentials. J. Neurophysiol. **107**(5), 1337–1355 (2012)
4. Chapin, J.K., Moxon, K.A., Markowitz, R.S., Nicolelis, M.A.: Real-time control of a robot arm using simultaneously recorded neurons in the motor cortex. Nat. Neurosci. **2**(7), 664–670 (1999)
5. Collinger, J.L., et al.: High-performance neuroprosthetic control by an individual with tetraplegia. Lancet **381**(9866), 557–564 (2013)
6. Devlin, J., Chang, M.W., Lee, K., Toutanova, K.: Bert: Pre-training of deep bidirectional transformers for language understanding. arXiv preprint arXiv:1810.04805 (2018)
7. Gilja, V., et al.: A high-performance neural prosthesis enabled by control algorithm design. Nat. Neurosci. **15**(12), 1752–1757 (2012)
8. Gilja, V., et al.: Clinical translation of a high-performance neural prosthesis. Nat. Med. **21**(10), 1142 (2015)
9. Hochberg, L.R., Bacher, D., Jarosiewicz, B., Masse, N.Y., Simeral, J.D., Vogel, J., Haddadin, S., Liu, J., Cash, S.S., Van Der Smagt, P., et al.: Reach and grasp by people with tetraplegia using a neurally controlled robotic arm. Nature **485**(7398), 372–375 (2012)
10. Jackson, A., Hall, T.M.: Decoding local field potentials for neural interfaces. IEEE Trans. Neural Syst. Rehabil. Eng. **25**(10), 1705–1714 (2016)
11. Li, Y., Qi, Y., Wang, Y., Wang, Y., Xu, K., Pan, G.: Robust neural decoding by kernel regression with Siamese representation learning. J. Neural Eng. **18**(5), 056062 (2021)
12. Liu, Q., et al.: Efficient representations of EEG signals for SSVEP frequency recognition based on deep multiset CCA. Neurocomputing **378**, 36–44 (2020)
13. Lu, J., Batra, D., Parikh, D., Lee, S.: ViLBERT: Pretraining task-agnostic visiolinguistic representations for vision-and-language tasks. arXiv preprint arXiv:1908.02265 (2019)
14. Pandarinath, C., et al.: High performance communication by people with paralysis using an intracortical brain-computer interface. Elife **6**, e18554 (2017)
15. Qi, Y., et al.: Dynamic ensemble bayesian filter for robust control of a human brain-machine interface. IEEE Trans. Biomed. Eng., 1–11 (2022). https://doi.org/10.1109/TBME.2022.3182588
16. Rickert, J., de Oliveira, S.C., Vaadia, E., Aertsen, A., Rotter, S., Mehring, C.: Encoding of movement direction in different frequency ranges of motor cortical local field potentials. J. Neurosci. **25**(39), 8815–8824 (2005)
17. Serruya, M.D., Hatsopoulos, N.G., Paninski, L., Fellows, M.R., Donoghue, J.P.: Instant neural control of a movement signal. Nature **416**(6877), 141–142 (2002)

18. Shi, Z., Chen, X., Zhao, C., He, H., Stuphorn, V., Wu, D.: Multi-view broad learning system for primate oculomotor decision decoding. IEEE Trans. Neural Syst. Rehabil. Eng. **28**(9), 1908–1920 (2020)
19. So, K., Dangi, S., Orsborn, A.L., Gastpar, M.C., Carmena, J.M.: Subject-specific modulation of local field potential spectral power during brain-machine interface control in primates. J. Neural Eng. **11**(2), 026002 (2014)
20. Taylor, D.M., Tillery, S.I.H., Schwartz, A.B.: Direct cortical control of 3D neuroprosthetic devices. Science **296**(5574), 1829–1832 (2002)
21. Vaswani, A., et al.: Attention is all you need. In: Advances in Neural Information Processing Systems, pp. 5998–6008 (2017)
22. Wang, W., Arora, R., Livescu, K., Bilmes, J.: On deep multi-view representation learning. In: International Conference on Machine Learning, pp. 1083–1092. PMLR (2015)
23. Wang, Y., Lin, K., Qi, Y., Lian, Q., Feng, S., Wu, Z., Pan, G.: Estimating brain connectivity with varying-length time lags using a recurrent neural network. IEEE Trans. Biomed. Eng. **65**(9), 1953–1963 (2018)
24. Willett, F.R., et al.: Hand knob area of premotor cortex represents the whole body in a compositional way. Cell **181**(2), 396–409 (2020)

A Mask Image Recognition Attention Network Supervised by Eye Movement

Rongkai Zhang, Libin Hou, Runnan Lu, Linyuan Wang, Li Tong, Ying Zeng, and Bin Yan(✉)

Henan Key Laboratory of Imaging and Intelligent Processing, PLA Strategy Support Force Information Engineering University, Zhengzhou 450001, China
yingzeng@uestc.edu.cn, ybspace@hotmail.com

Abstract. Mask image modeling has become a hot topic in the field of computer vision, which is widely used in the Image recognition of semantic breakage. The interference in the long-distance observation leads to the low ISAR (Inverse Synthetic Aperture Radar) imaging quality for aerospace vehicles. Mask image can simulate and close to the real ISAR image, and adequate simulation mask images can improve the robustness of the classifier. However, the classification performance of DNN for ISAR images is insufficient, especially for the mask images with high interference and high masking rate. To improve the machine recognition ability with the help of human brain recognition process, we propose Gaze Attention Supervise Network, which the fixation information is applied to assist in generating the attention of the network feature layer by attention supervised module. We apply eye-tracking label in mask image dataset as the supervision of network training. The proposed network fully learns the eye gaze region information to generate attention view. The results showed that classification performance improved 10.62% ~ 15.22%, especially in the small training dataset with high masking rate.

Keywords: Mask image recognition · Simulated ISAR · Eye-tracking label · Attention supervised module

1 Introduction

ISAR (Inverse Synthetic Aperture Radar) image is widely used in aerospace field and becomes an important imaging method for field observation [1]. Due to the long observation distance, limited observation position and strong signal interference, ISAR imaging quality is low and blurred [2].

Due to the limitation of data sources and military application background, it is difficult to obtain a large number of public ISAR images of real targets. Simulation pictures are generally used to provide samples needed to study object recognition methods [3]. In addition, adding masks to images can simulate low-quality ISAR images. For this

R. Zhang, L. Hou, R. Lu — These authors contributed to the work equally and should be regarded as co-first authors.

X. Ying (Ed.): HBAI 2022, CCIS 1692, pp. 30–38, 2023.
https://doi.org/10.1007/978-981-19-8222-4_3

type of mask image, masking autoencoder network (MAE) performs as a state of art approach in image classification, target detection, image segmentation and other conventional tasks. Using the sparse distribution of natural image information, autoencoder network is applied to mask images to help improve performance [4–6]. However, in the face of high masking rate ISAR image recognition, MAE cannot capture key features effectively. The machine has poor recognition performance for high masking rate ISAR images, which leads to the widespread application of manual classification.

For high masking rate images, the eye gaze area is modulated by the brain advanced cognitive, which has the advantages of fast, feelingless labeling and convenient. Therefore, we propose a mask image recognition attention network based on eye movement supervision, which called Gaze Attention Supervise Network (GASNet). In the proposed network, eye movement information is used to supervise and guide the training of network attention area, which can effectively combine with MAE. The fusion of eye movement information and deep neural network realizes information integration between machine learning and human cognition.

In our work, the eye gaze area is collected and analyzed during the aircraft identification tasks. Then, the attention information of eye movement was added into the network, and the eye gaze area was used to assist the training of the network attention. Our research shows that human brain intelligence and machine intelligence are integrated and complementary in network training, which improves the accuracy and efficiency of network training.

The main contributions of this paper are as follows:

- The proposed GASNet with eye track supervision. GASNet has prominent effect in high masking rate mask image recognition, and has stable improvement in various classification backbone networks. Besides, no eye data input is required in the test phase after the network obtains fixation ability.
- Gaze attention Consistence training Module (GC Module) is presented in this framework, which supervise the generation of network attention map while classification training. The network attention of this module is closer to the eyetrack compared with the traditional method.

2 Methods

GASNet was proposed to align neural network attention with external supervision of eye tracking visual attention, as shown in Fig. 1. Firstly, eyetrack collecting module gets the eye movement fixation lables, and gaze map A is generated after processing. Meanwhile, classification training module trains the classification backbone model, and feature map G is extracted from fully connected layers. Finally, the attention of eye track supervised classification network training through loss consistency between map G and map A in GC Module.

2.1 Datasets

Simulated ISAR mask image dataset (SimISAR) included a total of 3801 ISAR dynamic videos (7 targets, 181 angles, 3 random modes).The original experimental stimulus

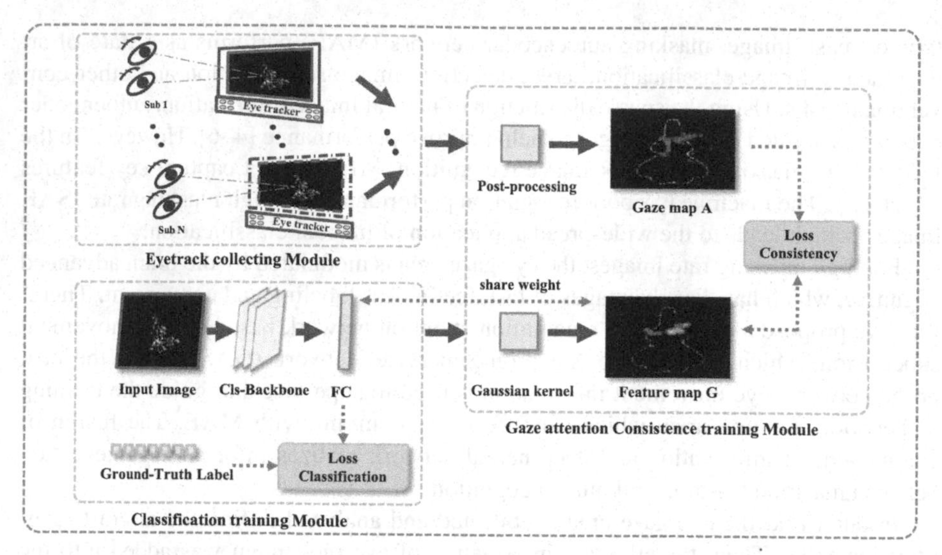

Fig. 1. Overview of our Gaze Attention Supervise Network (GASNet) framework. Eyetrack collecting Module gets the eye movement fixation lables. Classification training Module trains the classification backbone model. Gaze attention Consistence training Module (GC Module) supervises the generation of Feature map G with Gaze map A.

material was the public ISAR simulation data set SynISAR [7]. SimISAR was produced from SynISAR through the following processing. Further processing including: image amplification, pseudocolor filling, image masked, dynamic video played (masking level from high to low) [8]. In the eye tracking experiment, the SPECTRUM equipment of Tobii company was used to collect the eye movement data of 18 subjects watching random 273 (39 pictures per target) videos and judging the target type in each video.

2.2 The Generation of Gaze Heat Map

The generation of heat map includes two stages: data cleaning and preprocessing. In the data cleaning stage, we used the speed threshold recognition gaze points classification algorithm in previous studies [9] to remove the noise. In the preprocessing stage, the filtered data were firstly segmented according to the type and rotation angle of the target, and the data segments with more invalid data were eliminated. Secondly, eliminate the gaze points falling outside the screen, which the coordinates were greater than the screen edge, and fill the blank data widow caused by eliminating invalid data. Thirdly, unify the angles of different angles of the same type target through central rotation. Fourthly, subtract the baseline value of the global average of all trials according to the category, trials number and long-term weighting. Finally, a 35x35 size Gaussian kernel was applied to smooth the gazing point and obtain the eye movement heat map [10].

2.3 Network Architecture

In order to obtain the effectiveness of attention consistency, we used fashion classification network backbones to establish an eye tracking attention training framework. Therefore, the GASNet was introduced in terms of classification backbone and supervision module respectively. The process of GASNet is shown in Algorithm 1.

Algorithm 1 The process of GASNet

Input: The labeled SimISAR dataset (*s* is the number of training sets, *e* is the training epoch) and eye track data

Output: the model parameter of GASNet and the predicted labels.

1: **for** i=1,…,s and k=1,…,e **do**:

2: Train classification module via *loss* L_{cls} ;

3: Generate Gaze map A from eye track data via Post-processing;

4: Obtain the feature map from FC of module in step1;

5: Generate Feature map G from feature map in step 3 via Gaussian kernel;

6: Process consistency between map G and map A via *loss* L_{mse}

7: Share the weight from map G to FC in step3

8: **end for**

9: Obtain the model parameter of GASNet and generate the predicted labels via GASNet.

Classification Backbone. Firstly, we selected the popular classification network including ResNet [11], Swin-Transformer [12] and MAE [13] as the backbone of our task, all of which inherit the pretrained weights on ImageNet. The output characteristics from the backbone network were mapped and then processed through full connected layers (FC) to produce the required 7-class outputs. We used the cross-entropy loss as classification loss, which was calculated between GASNet predicted labels and the real ISAR aircraft label.

Gaze Attention Consistence Training Module. GC Module was taken as a part of training module to adapt to the supervision of the eye tracking attention, as shown in Fig. 1. The mean square error (MSE) between the two attentional forces was taken as the objective function to achieve the purpose of monitoring attention network training by eye movement fixation. Map A was the fixation attentional diagram obtained by eye movement, and map G was the network attentional diagram processed by neural network, then the consistency loss of MSE can be expressed as:

$$L_{mse} = \sum_{x,y} (A(x, y) - G(x, y))^2 \tag{1}$$

A series of data cleaning processes were carried out on the eye-movement trajectory graph to adapt to the strict constraint in the preprocessing part. This MSE constraint strictly required the consistency between the network and the attentional graph.

3 Results

3.1 Eye Movement Heat Map

Figure 2 shows the overlay eye movement heat map of all subjects when they watched target 1, target 2 and target 3. From the comparison of the horizontal axis, it can be seen that when looking at different aircraft targets, the eye movements show different gaze trajectories, which can be seen as the brain extracts different features for different target types. We found that for the three types of targets, the eye movement trajectory was mainly concentrated in the upper half of the aircraft. The reason may be that there was a high similarity between the three targets, while the difference was mainly concentrated in the upper half of the aircraft (with or without duck wings). From the comparison of

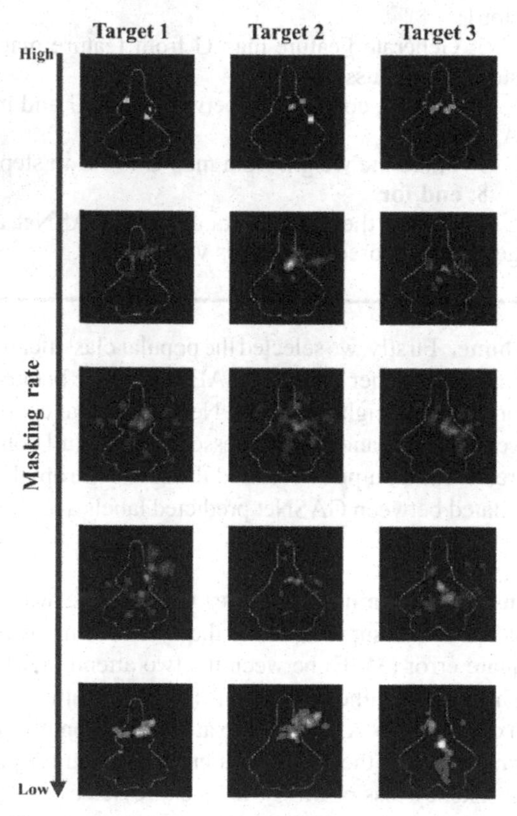

Fig. 2. Eye movement heat map of three targets at different masking rates. The arrow represents masking rate decreases from high to low.

longitudinal axis (high masking rate to low masking rate), the difference was transferred to the lower part of the aircraft. Therefore, the subjects may firstly make a preliminary distinction between the three targets through this main difference at the beginning of the video playback, and collect more information in the later stage of video playback to determine the target type. This phenomenon reflects the unique strategy of human beings in target classification.

3.2 Network Performance

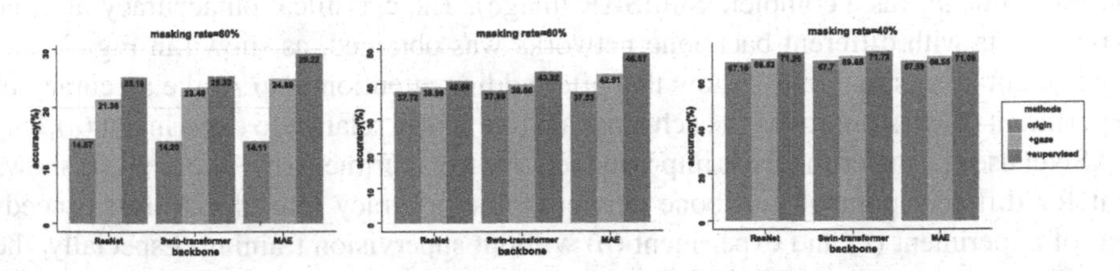

Fig. 3. Further comparison on classification accuracy (%) of the experiments on high masking rate. All backbone networks gain high enhance of accuracy on the lowest expose rate, especially MAE.

We designed the following experiments for comparison, and test performance of the network is shown in Fig. 3.

a) Classification by backbone network with mask image as input only.
b) The attention map A is added to the mask image as input and classified through the backbone network.
c) Supervisory training with attention map A combined with network attention map.

Table 1. Classification accuracy (%) of the experiments on different backbone networks with variable masking rate.

Masking rate	80%	60%	40%	20%	0%
ResNet	14.57	37.72	67.15	92.17	98.55
+ gaze	21.36	38.99	68.53	93.24	99.67
+ supervised	25.19	40.66	71.20	93.44	**99.97**
Swin-transf	14.20	37.99	67.70	92.62	98.19
+ gaze	23.49	39.66	69.65	92.91	98.64
+ supervised	25.32	43.22	**71.72**	94.50	98.70

(continued)

Table 1. *(continued)*

Masking rate	80%	60%	40%	20%	0%
MAE	14.11	37.53	67.59	92.71	98.96
+ gaze	24.89	42.51	69.55	94.85	99.63
+ supervised	**29.22**	**48.57**	71.09	**94.72**	99.06

The images with different masking rate were used to train (When masking rate is 0%, the input image was a complete SimISAR image). The classification accuracy of three experiments with different backbone networks was obtained, as shown in Fig. 3. The experiment results in Table 1 show that after adding attention map A, the ac-curacy of experiment (b) was improved in each masking rate image relative to experiment (a). The GASNet used by experiment c is improved relative to experiment (b). The analysis shows that for different popular backbone networks, the accuracy rate of GASNet exceeds that of experiment (a) and experiment (b) without supervision training. Especially, the classification accuracy of MAE is lower than ResNet and Swin-transformer when the masking rate is high. But the improvement is the largest after adding the supervision module, and the accuracy exceeds the other two backbone networks after adding the supervision module. The experiment results show that the effectiveness of GASNet using attention map and reflect the potential of MAE in processing masked images.

3.3 Network Attention Visualization

As shown in Fig. 4, two representative SimISAR aircraft images are taken as the example to visualize the images at each experiment stage. The Fig. 4 (a) (b) (c) show the process from the original image to high mask processing and finally to otain the eye gaze area. The Fig. 4 (d) (e) (f) reveal the changes of network attention area by eye heat map assisted training.

Network attention with eye-tracking supervised effectively learned the region of human eyes interest. In the Fig. 4 (f1), we find that the network attention successfully focused on the aircraft wing, which is very close to the eye gaze in Fig. 4 (c1). The proposed network can locate areas close to real human eye attention. In addition, the subject's questionnaire generally shows that the wing was the focus area in the classification task, which was also the most significant area to compare various types of aircraft.

Compared with the original attention network, the proposed network shrinks the range of attention. Figure 4 (f1) combines Fig. 4 (b1) eye gaze area with Fig. 4 (d1) and Fig. 4 (e1) machine attention area. We speculate that the improvement of classification performance was due to the attention of eye movement annotation supervision network was more focused on the aircraft difference. Smoothing was applied in previous thermal mapping to reduce the interference of scattered eye saccade, which help the network focus on high perceived value positions.

The slender aircraft is shown in the Fig. 4 right column. We can see the aircraft attention difference between human eyes and machines. In Fig. 4 (c2), the eye gaze area

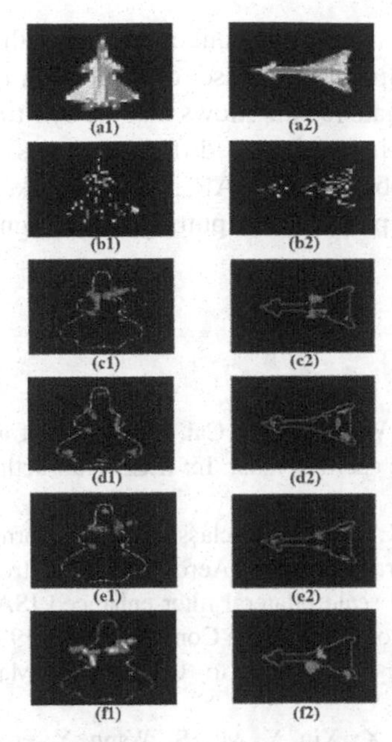

Fig. 4. SimISAR image display and network attention graph visualization. (a) Original SimISAR image. (b) SimISAR image with high mask rate. (c) Global average eye gaze. (d) GASNet feature maps without eye gaze supervised. (e) GASNet feature maps with eye gaze input. (f) GASNet attention map with eye gaze supervised.

is concentrated in the middle of the aircraft, and the classification can be distinguished only by looking at the thickness of the aircraft middle position. However, the gaze area of the backbone network is more dispersed, and the attention is evenly distributed at both sides of the aircraft. Figure 4 (f2) shows the attention graph of the proposed network. The network gaze area not only focuses on the middle part of the aircraft, but also increases the attention at the tail of the aircraft. The network supervised by eye-tracking combines the attention of human eyes and machines.

The proposed network still had a focus area for ISAR images with high mask rate. We speculate that locating the difference position helps to improve performance when the image was highly blurred. The supervised training of eye labeling help the network learn the human brain characteristics of noise suppression and high robustness. Eye movement information helps network attention predict what people are interested in.

4 Conclusion

GASNet is proposed in this paper, a deep learning masked image classification framework based on eye movement attention supervision. In this method, the GC Module is presented, which supervise the generation of network attention map while classification training. We use the eye track heat map of masked image containing human brain advanced modulated attention to improve neural network performance. Experiments show that GASNet achieved an improvement compared with the mainstream classification network backbones in SimISAR dataset classification performance. Furthermore, the visualization of intermediate results shows that the eye-tracking supervised network has more focused attention in the image difference areas. This method provides an efficient solution for the problem that ISAR is often masked and difficult to identify. Moreover, a new pipeline is provided for putting brain cognitive into computer vision process system.

References

1. Cheng, B., Ge, J., Cheng, W., Chen, Y., Cai, Y., Qi, C., et al. Real-time imaging with a 140 GHz In-verse synthetic aperture radar. In: IEEE Transactions on Terahertz Science and Technology 3, pp. 594–605
2. Lee, S.J., Park, S.H., Kim, K.T.: Improved classification performance using ISAR images and trace transform. In: IEEE Transactions on Aerospace & Electronic Systems 53, pp. 950–965
3. Bozkurt, H., Erer, I.: Multi scale bilateral filter enhanced ISAR target recognition. Signal Processing & Communication Application Conference, pp. 1993–1996
4. Cao, S., Xu, P., Clifton, D.A.: How to Under-stand Masked Autoencoders. ArXiv abs/2202.03670 (2022)
5. Chen, X., Ding, M., Wang, X., Xin, Y., Mo, S., Wang, Y., et al.: Context Autoencoder for Self-Supervised Representation Learning. ArXiv abs/2202.03026 (2022)
6. Xie, Z., Zhang, Z., Cao, Y., Lin, Y., Bao, J., Yao, Z., et al.: SimMIM: A Simple Framework for Masked Image Modeling. ArXiv abs/2111.09886 (2021)
7. Kondaveeti, H.K.: SynISAR. International Con-ference on Emerging Trends in Engineering, Technology and Science (ICETETS-2016), India. Zenodo
8. Stember, J.N., Celik, H., Krupinski, E., Chang, P.D., Bagci, U.: Eye tracking for deep learning segmentation using convolutional neural networks. Journal of Digital Imaging
9. Salvucci, D.D., Goldberg, J.H.: Identifying fixations and saccades in eye-tracking protocols. In: Proceedings of the Eye Tracking Research & Application Symposium, ETRA 2000, Palm Beach Gardens, Florida, USA, November 6–8 (2000)
10. Wooding, D.S.: Fixation maps: Quantifying eye-movement traces. Proceedings of the Eye Tracking Research & Application Symposium, ETRA 2002, New Orleans, Louisiana, USA, March 25–27 (2002)
11. He, K., Zhang, X., Ren, S., Sun, J.: Deep re-sidual learning for image recognition. IEEE Conference on Computer Vision and Pat-tern Recognition (CVPR) **2016**, 770–778 (2016)
12. Liu, Z., Lin, Y., Cao, Y., Hu, H., Wei, Y., Zhang, Z., et al.: Swin transformer: hierarchical vision transformer using shifted windows. In: 2021 IEEE/CVF International Conference on Computer Vision (ICCV), pp. 9992-10002 (2021)
13. He, K., Chen, X., Xie, S., Li, Y., Dollár, P., Girshick, R.: Masked Autoencoders Are Scalable Vision Learners. ArXiv abs/2111.06377 (2021)

DFC-SNN: A New Approach for the Recognition of Brain States by Fusing Brain Dynamics and Spiking Neural Network

Yan Cui[1,2,3], Wuque Cai[3], Tianyao Long[3], Hongze Sun[3], Dezhong Yao[3], and Daqing Guo[3(✉)]

[1] Department of Neurosurgery, Sichuan Provincial People's Hospital, University of Electronic Science and Technology of China, Chengdu 610072, China

[2] Chinese Academy of Sciences, Sichuan Translational Medicine Research Hospital, Chengdu 610072, China

[3] The Clinical Hospital of Chengdu Brain Science Institute, MOE Key Lab for NeuroInformation, Centre for Information in Medicine, School of Life Science and Technology, University of Electronic Science and Technology Of China, Chengdu 611731, China

202011140121@std.uestc.edu.cn, sunhongze@aliyun.com, {dyao,dqguo}@uestc.edu.cn

Abstract. Rich dynamics are the intrinsic features in brain activity, which could be characterized as sequences of multiple spatio-temporal activity events. However, how to efficiently apply brain dynamics for the recognition of brain states is still unclear and need more investigations. The spiking neural network (SNN) is a promising model with better performance in the pattern recognition of event streams. Thus, this paper proposes an algorithm framework for brain states recognition by fusing brain dynamics and SNN, where the brain dynamics are estimated as the dynamic functional connectivity (DFC) matrices. Through applying the DFC-SNN algorithm to the dataset of resting state electroencephalograph signals from healthy subjects and obsessive compulsive disorder patients, we observed that this algorithm was competent to perform the recognition of pathological brain states. It showed that the convergence of SNN model was rapid within less than 20 epochs, and the accuracy was 87.5% under optimal threshold of DFC matrices. In summary, this is the first attempting for the recognition of brain states via the aspect of brain dynamics. The algorithm framework would be beneficial for the applications of SNN model in the field of neuroscience.

Keywords: Spiking neural network · Dynamic functional connectivity · Brain states

Supported by the Hospital Fund of Sichuan Provincial People's Hospital (Grant No. S2022QN0132), the MOST 2030 Brain Project (Grant No. 2022ZD0208500) and the CAMS Innovation Fund for Medical Sciences (CIFMS) (Grant No. 2019-I2M-5-039).

X. Ying (Ed.): HBAI 2022, CCIS 1692, pp. 39–49, 2023.
https://doi.org/10.1007/978-981-19-8222-4_4

1 Introduction

As the third-generation artificial neural network, the spiking neural network (SNN) is believed to have originated in the neural mechanism of brain activity [28,34]. SNN is a brain-inspired neural network model, which could extract the spatio-temporal features from the event streams [21]. Moreover, because of the abundant neural dynamic characters in spatio-temporal domain, multi-encoding mechanisms and event-driven advantages, SNN has called into more attention in the field of artificial intelligence [19,20,39]. Compared with traditional neural network models, such as the recurrent neural network (RNN) [38] and conventional neural network (CNN) [27], SNN provided better solution to the event-based data processing and has stronger computing power [33]. Previous work has demonstrated that SNN showed better performance in human gesture recognition [2], object tracking [13] and the recognition of human postures [15] and event-based hand gesture [37]. However, the applications of SNN in the processing of neuroimaging signals is still fewer. To our knowledge, there is only one investigation combining SNN with brain neural activity, which employed the SNN model for the detection of high frequency oscillations in the intraoperative electrocorticography recordings [6]. Interestingly, this study furtherly illustrated the potential roles of SNN model in the classification of specific brain states. Thus, how to effectively fuse the SNN model with brain activity is reasonable for more investigations.

Our brain displays inherent dynamic activity in integrating both internal and external stimuli [5,16,25]. The dynamics of brain activity could be characterized by the transitions of quasi-stable brain states with several promising dynamic analytic methods, including the dynamic functional connectivity method [14], the microstate method [23] and the coactive micropattern method [8]. Accumulating evidence has indicated that corresponding to the resting state, the pathological state and the task state, there exist specific structures of brain dynamics [4,31]. It showed that during the resting state, the brain dynamics displayed local integration and modular topology, accompanied with scale free and heavy tail properties [10,26]. Moreover, in different conscious states, such as the wakefulness-sleep cycle, the dynamic jumps of brain states are also distinct and showed high dependence to the levels of consciousness [32]. Besides, during movie viewing the brain dynamics became a temporal sequence of functional states, while it is a predominantly transitions within two distinct brain states during the resting state [22]. In the studies of brain diseases, the clinical relevance of the brain dynamic structure and its potential biomarker utility have also been gradually proposed and received wide attention. It has been reported that the integration of brain dynamics reduced in schizophrenic patients, and their brains preferred the states with lower connectivity [11]. In absence epilepsy patients' brain, there was an abnormal complex conversion structure in brain dynamics when suffering seizures, suggesting that adaptive remodeling may occur in the brain after seizures [18]. Taken together, all these works revealed the unique brain dynamics of different brain states, and indicated that the temporal transitions might be the underlying neural mechanism for them. However, there still lack further investigations of the brain dynamics in classifying brain states.

Accordingly, we addressed this question by novelly combining brain dynamics with the SNN model in the present work. Based on the findings about the specificity of brain dynamics for different brain states, we hypothesized that these specificities were capable for the discrimination of brain states.

Therefore, we firstly estimated the brain dynamics with the dynamic functional connectivity (DFC) method and considered the brain dynamics as event streams, where each discrete brain functional connectivity map was an independent event. Then we trained the brain dynamics with the SNN model to find the best network model in distinguishing different brain states. By employing the DFC-SNN algorithm to the dataset containing the resting state brain activity of both healthy subjects and obsessive compulsive disorder (OCD) patients, we found that the accuracy could be 87.5%. This is the first attempt of fusing brain dynamics and SNN model, which showed significant and valuable potential in further works about the classification in neuroscience.

2 Methods

2.1 DFC-SNN Framework

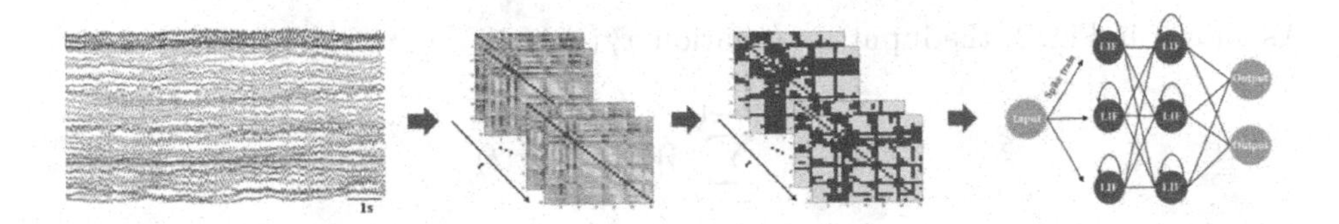

Fig. 1. The framework of DFC-SNN algorithm.

In the present work, we combined the dynamic functional connectivity (DFC) and the SNN structure, and proposed a new algorithm for the classification of neural signals. The new algorithm was named the DFC-SNN algorithm, which contained several steps, as showed in Fig. 1.

- Step 1: Estimate the dynamic functional connectivity matrices from the original EEG signals, with the sliding window method.
- Step 2: Threshold the dynamic functional matrix to get the binary matrices, and create the dynamic dataset.
- Step 3: Train and test the dynamic dataset with the spike neural network.

SNN is composed of spiking the leaky integrate-and-fire (LIF) neuron, which is governed by:

$$\tau_m \frac{dV}{dt} = -(V - V_r) + x, \tag{1}$$

$$o = \begin{cases} 1 & \text{if } V \geq V_{\text{th}} \\ 0 & \text{else,} \end{cases} \tag{2}$$

where V is the membrane potential, V_{r} is the resting potential, x is the external input current and τ_m is the membrane time constant. Note that we set V_{r} as 0 in the current study. Besides, when the membrane potential exceeds the threshold (V_{th}), the neuron would be activated. o is the activated state of the neuron, where 1 represents activation and 0 is deactivation. LIF neuron will fire a spike and reset the membrane potential to V_{r}. As shown in Fig. 2, at single-neuron level, an LIF neuron receives information from the preceeding layers in the spatial domain (SD), and inherents the history memory in the temporal domain (TD) simultaneously. For simplicity, the discreted form of Eq. 1 and Eq. 2 used in network is given by:

$$V_i^n(t) = \beta_i^n(t) \cdot V_i^n(t-1) + \frac{dt}{\tau_m} \cdot x_i^n(t), \tag{3}$$

where $V_i^n(t)$ represents the membrane potential of the i-th neuron in the n-th layer at time t. Theoretically, the decay factor of membrane potential $\beta_i^n(t)$ can be described as followed:

$$\beta_i^n(t) = \begin{cases} 1 - \frac{dt}{\tau_m} & \text{if } o^{t-1} = 0 \\ 0 & \text{if } o^{t-1} = 1. \end{cases} \tag{4}$$

As shown in Fig. 2, the input information $x_i^n(t)$ is:

$$x_i^n(t) = \sum_{j=1}^{l(n-1)} W_{i,j}^n \cdot o_j^{n-1}(t), \tag{5}$$

where the synaptic weights between the pre- and post- neuron is denoted as $W_{i,j}^n$. To evaluate the distance between the labels of sample and the outputs of SNN model, the mean square error (MSE) parameter is used in our work:

$$L_{mse} = \frac{1}{2S} \sum_{s=1}^{S} \left[y_s - \frac{1}{T} \sum_{t=1}^{T} o_s^N(t) \right] \tag{6}$$

where N and T is the number of layers and the length of time-window in our SNN model, respectively. The S represents the number of samples feeded into SNN. The feedforward path of spatio-temporal information in our SNN is established by Eqs. (3)–(6). To cope with the non-differentiable property of spiking events, we use the spatio-temporal backpropagation algorithm to train our SNN model directly based on supervised learning [7,35].

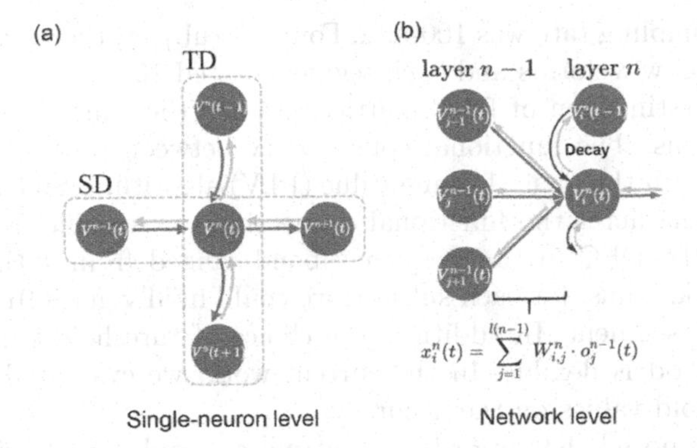

Fig. 2. The SNN model. (a) In the single-neuron level. (b) In the network level.

Besides, we established an all-connected SNN model. The details of this all-connected SNN model was described as followed. The SNN model had four layers, including the input layer, two hidden layers and the output layer. Specifically, the topology of SNN is [input-1024FC-1024FC-output]. In addition, during the learning of SNN, we applied the index decay strategy (hyper-parameter: gamma) to character the decrease of learning rate. The other hyper-parameters of SNN in the current work are displayed in Table 1.Specifically, we set the threshold potential V_{th} as 0.2 mV, the time constant for membrane potential τ_m as 2.5 ms. In addition, the batch size in the present model is 32, the initial learning rate is set to 0.001 and the gamma value is set to 0.93.

Table 1. Hyper-parameters setting.

Hyper-parameters	Values
V_{th}	0.2 mV
τ_m	2.5 ms
Batch size	32
Learning rate	0.001
Gamma	0.93

2.2 Dataset

In the present work, we employed the dataset in our prior work to evaluate the performance of DFC-SNN algorithm [17]. Briefly, there were two categories of resting state electroencephalogram (EEG) patterns in this dataset, including 36 healthy subjects and 50 obsessive compulsive disorder (OCD) patients. The resting state EEG signals were acquired from 30 brain regions according to the 10/20 systems, through the Neuroscan Nuampls digital amplifier system. In addition, the sampling rate was 1000 Hz. For each subject, there were 4 segments with stable EEG waveforms and each segment lasted 10 s.

During the estimation of DFC matrices, we set the window as 1 s, and the overlap as 100 ms. The functional connectivity between pairs of brain regions was estimated with the phase locking value (PLV) algorithm. Since that the PLV method could character the functional connectivity in specific frequency band, we calculated the DFC in the frequency band ranged from 4 Hz to 13 Hz in the present work. Thus, for each subject, we could finally get a $91 * 30 * 30$ DFC matrix for each segment. In addition, the choice of threshold value during the DFC-SNN method is flexible. In the current work, we evaluated the effects of different threshold values on the accuracy.

It should be noted that in order to get more samples, we divided the DFC matrix for each subjects into three matrices with same sizes. Thus, we finally had 432 samples for the healthy subjects' group, and 600 samples for the OCD patients' group. Each sample had a $30 * 30 * 30$ matrix descripting the dynamics of brain network connectivity for every subject.

During the evaluation of DFC-SNN performance, we firstly trained the SNN model with 832 samples randomly selected from the whole 1032 samples. In the present SNN model, the output layer contained two neurons, which represented the two categories of EEG patterns, respectively. Then, we tested the model with the other 200 samples, and defined the accuracy of recognition as the percentage of corrected predictions in these samples. Besides, in order to find the better and stable performance of SNN model, we trained the SNN model for 200 epochs and considered the highest accuracy as the classification accuracy in specific condition.

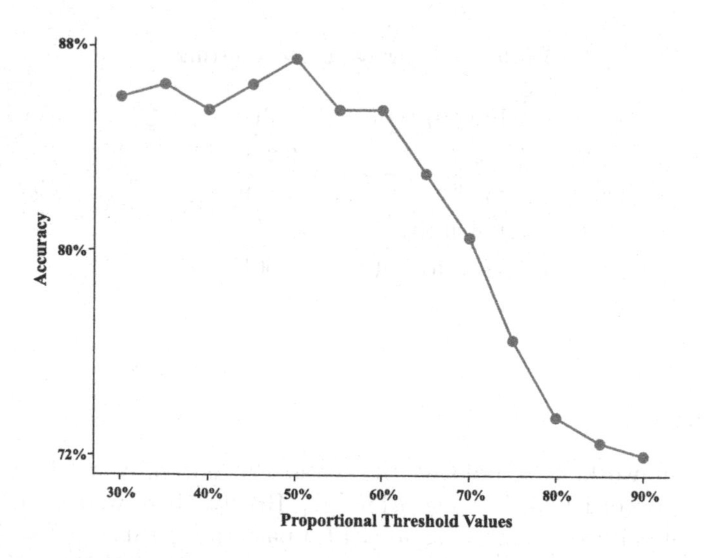

Fig. 3. The classification accuracy under different proportional threshold values.

3 Results

In the DFC-SNN algorithm, thresholding the dynamic functional connectivity matrices to binary matrices is a very important step and have direct effects on the final performance of classification. To address the effect of threshold value choice, we firstly evaluated the performance of DFC-SNN method under different threshold values. In the current work, we used the proportional thresholding method to get the binary matrix. This method could preserve a proportion p ($0<p<1$) of the strongest functional connectivities and set the other connections as 0. In the present work, we evaluated different proportional threshold values ranged from 30% to 90% with the step as 5%.

As shown in Fig. 3, we observed that the classification accuracies increased with the smaller proportional threshold values (less than 50%). When the proportional threshold values were larger than 50%, the classification accuracies decreased with the increase of proportional threshold values. The maximum classification accuracy emerged and was 87.5% when the proportional threshold value was 50%. This phenomenon might demonstrate that half of the strongest connectivities among brain regions could effectively reflect the dynamics of brain activity and reveal the characterization of brain dynamics.

In addition, we found that when the proportional threshold values ranged from 30% to 70%, the classification accuracies all exceed 80%, suggesting that the DFC-SNN algorithm was capable for the classification of brain neural signals during the pathological state.

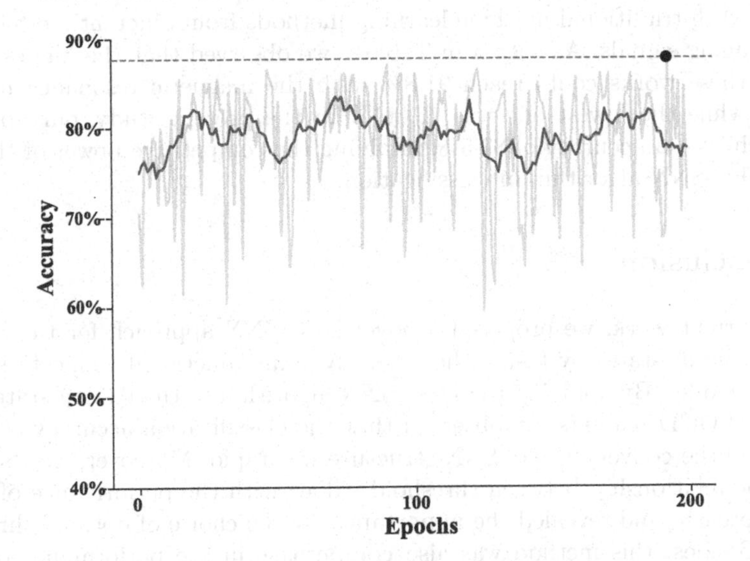

Fig. 4. The classification accuracy under proportional threshold value 50%. The nattier blue line showed original accuracy values in 200 epochs and the blue line is the 10-epochs moving average.

Secondly, we furtherly investigated the variability of classification accuracy under 200 training epochs when the proportional threshold value was 50%. As showed in Fig. 4, we found that the classification accuracies displayed relatively stable variability among 200 training epochs. Moreover, we observed a rapid convergence of SNN model within less than 20 training epochs. By considering the capability of SNNs in spatio-temporal information processing, it has potential to show better performance in the classification tasks of brain states.

Table 2. Literature of prior classification works of OCD patients.

Papers	Samples	Neuroimaging signals	Accuracy
[24]	86	Magnetic resonance imaging	73.86%
[3]	74	Electroencephalogram	85.00%
[12]	79	Electroencephalogram	81.04%
[30]	29	Magnetic resonance imaging	80.00%
[36]	128	Magnetic resonance imaging	91.80%
[29]	26	Functional Magnetic resonance imaging	89.00%
[1]	57	Electroencephalogram	89.66%
[9]	78	Magnetic resonance imaging	87.50%

Besides, we furtherly tracked prior works on the classification of OCD patients with traditional machine learning methods from different modalities of neuroimaging signals. As shown in Table 2, we observed that the highest accuracy in these works could reach 91.8% with the magnetic resonance imaging signals, while the lowest one was 73.86%. In the present study, our approach could achieve an accuracy of 87.5%, implying the competitive power of the proposed DFC-SNN algorithm in classification.

4 Conclusion

In the current work, we proposed a novel DFC-SNN approach for the classification of brain states by fusing the brain dynamic functional connectivity and the SNN model. By applying the DFC-SNN algorithm to the resting state brain activity of OCD patients, we observed that the classification accuracy could be 87.5% and the convergence of SNN structure was rapid. Moreover, we also evaluated the relationship between threshold values with the performance of DFC-SNN approach, and revealed the importance for the choice of optimal threshold values. Besides, this method was also competitive in the performance of classification when compared with those in prior studies. Overall, our work was a new attempt in artificial intelligence via the dynamic structure of brain activity, which might bring some enlightenment in further works.

References

1. Altuğlu, T.B., et al.: Prediction of treatment resistance in obsessive compulsive disorder patients based on EEG complexity as a biomarker. Clin. Neurophysiol. **131**(3), 716–724 (2020)
2. Amir, A., et al.: A low power, fully event-based gesture recognition system. In: Proceedings of the IEEE Conference on Computer Vision and Pattern Recognition, pp. 7243–7252 (2017)
3. Aydin, S., Arica, N., Ergul, E., Tan, O.: Classification of obsessive compulsive disorder by EEG complexity and hemispheric dependency measurements. Int. J. Neural Syst. **25**(03), 1550010 (2015)
4. Braun, U., Schaefer, A., Betzel, R.F., Tost, H., Meyer-Lindenberg, A., Bassett, D.S.: From maps to multi-dimensional network mechanisms of mental disorders. Neuron **97**(1), 14–31 (2018)
5. Breakspear, M.: Dynamic models of large-scale brain activity. Nat. Neurosci. **20**(3), 340–352 (2017)
6. Burelo, K., Sharifshazileh, M., Krayenbühl, N., Ramantani, G., Indiveri, G., Sarnthein, J.: A spiking neural network (SNN) for detecting high frequency oscillations (HFOS) in the intraoperative ECOG. Sci. Rep. **11**(1), 1–10 (2021)
7. Cramer, B., et al.: Surrogate gradients for analog neuromorphic computing. Proc. Natl. Acad. Sci. **119**(4), e2109194119 (2022)
8. Cui, Y., et al.: Dynamic configuration of coactive micropatterns in the default mode network during wakefulness and sleep. Brain Connect. **11**(6), 471–482 (2021)
9. Cuicui, J., et al.: Disrupted asymmetry of inter-and intra-hemispheric functional connectivity at rest in medication-free obsessive-compulsive disorder. Front. Neurosci. **15**, 645 (2021)
10. Di, X., Biswal, B.B.: Dynamic brain functional connectivity modulated by resting-state networks. Brain Struct. Funct. **220**(1), 37–46 (2015)
11. Du, Y., et al.: Interaction among subsystems within default mode network diminished in schizophrenia patients: a dynamic connectivity approach. Schizophrenia Res. **170**(1), 55–65 (2016)
12. Erguzel, T.T., Ozekes, S., Sayar, G.H., Tan, O., Tarhan, N.: A hybrid artificial intelligence method to classify trichotillomania and obsessive compulsive disorder. Neurocomputing **161**, 220–228 (2015)
13. Hinz, G., et al.: Online multi-object tracking-by-clustering for intelligent transportation system with neuromorphic vision sensor. In: Kern-Isberner, Gabriele, Fürnkranz, Johannes, Thimm, Matthias (eds.) KI 2017. LNCS (LNAI), vol. 10505, pp. 142–154. Springer, Cham (2017). https://doi.org/10.1007/978-3-319-67190-1_11
14. Hutchison, R.M., et al.: Dynamic functional connectivity: promise, issues, and interpretations. Neuroimage **80**, 360–378 (2013)
15. Jiang, Z., et al.: Mixed frame-/event-driven fast pedestrian detection. In: 2019 International Conference on Robotics and Automation (ICRA), pp. 8332–8338. IEEE (2019)
16. Le Van Quyen, M., Bragin, A.: Analysis of dynamic brain oscillations: methodological advances. Trends Neurosci. **30**(7), 365–373 (2007)
17. Lei, H., Cui, Y., Fan, J., Zhang, X., Zhong, M., Yi, J., Cai, L., Yao, D., Zhu, X.: Abnormal small-world brain functional networks in obsessive-compulsive disorder patients with poor insight. J. Affect. Disord. **219**, 119–125 (2017)

18. Liao, W., et al.: Dynamical intrinsic functional architecture of the brain during absence seizures. Brain Struct. Funct. **219**(6), 2001–2015 (2014)
19. Liu, C., Shen, W., Zhang, L., Du, Y., Yuan, Z.: Spike neural network learning algorithm based on an evolutionary membrane algorithm. IEEE Access **9**, 17071–17082 (2021)
20. Lobo, J.L., Del Ser, J., Bifet, A., Kasabov, N.: Spiking neural networks and online learning: an overview and perspectives. Neural Netw. **121**, 88–100 (2020)
21. Maass, W.: Networks of spiking neurons: the third generation of neural network models. Neural Netw. **10**(9), 1659–1671 (1997)
22. Meer, J.N., Breakspear, M., Chang, L.J., Sonkusare, S., Cocchi, L.: Movie viewing elicits rich and reliable brain state dynamics. Nat. Commun. **11**(1), 1–14 (2020)
23. Michel, C.M., Koenig, T.: EEG microstates as a tool for studying the temporal dynamics of whole-brain neuronal networks: a review. Neuroimage **180**, 577–593 (2018)
24. Parrado-Hernández, E., et al.: Discovering brain regions relevant to obsessive-compulsive disorder identification through bagging and transduction. Med. Image Anal. **18**(3), 435–448 (2014)
25. Pesaran, B., et al.: Investigating large-scale brain dynamics using field potential recordings: analysis and interpretation. Nat. Neurosc. **21**(7), 903–919 (2018)
26. Ramirez-Mahaluf, J.P., et al.: Transitions between human functional brain networks reveal complex, cost-efficient and behaviorally-relevant temporal paths. NeuroImage **219**, 117027 (2020)
27. Razzak, M.I., Imran, M., Xu, G.: Efficient brain tumor segmentation with multi-scale two-pathway-group conventional neural networks. IEEE J. Biomed. Health Inform. **23**(5), 1911–1919 (2018)
28. Roy, K., Jaiswal, A., Panda, P.: Towards spike-based machine intelligence with neuromorphic computing. Nature **575**(7784), 607–617 (2019)
29. Sen, B., Bernstein, G.A., Mueller, B.A., Cullen, K.R., Parhi, K.K.: Sub-graph entropy based network approaches for classifying adolescent obsessive-compulsive disorder from resting-state functional MRI. NeuroImage: Clinic. **26**, 102208 (2020)
30. Sen, B., et al.: Classification of obsessive-compulsive disorder from resting-state FMRI. In: 2016 38th Annual International Conference of the IEEE Engineering in Medicine and Biology Society (EMBC), pp. 3606–3609. IEEE (2016)
31. Shine, J.M., Koyejo, O., Poldrack, R.A.: Temporal metastates are associated with differential patterns of time-resolved connectivity, network topology, and attention. Proc. Natl. Acad. Sci. **113**(35), 9888–9891 (2016)
32. Stevner, A., et al.: Discovery of key whole-brain transitions and dynamics during human wakefulness and non-rem sleep. Nat. Commun. **10**(1), 1–14 (2019)
33. Tavanaei, A., Ghodrati, M., Kheradpisheh, S.R., Masquelier, T., Maida, A.: Deep learning in spiking neural networks. Neural Netw. **111**, 47–63 (2019)
34. Wade, J.J., McDaid, L.J., Santos, J.A., Sayers, H.M.: Swat: a spiking neural network training algorithm for classification problems. IEEE Trans. Neural Netw. **21**(11), 1817–1830 (2010)
35. Wu, Y., Deng, L., Li, G., Zhu, J., Shi, L.: Spatio-temporal backpropagation for training high-performance spiking neural networks. Front. Neurosci. **12**, 331 (2018)
36. Xing, X., Jin, L., Shi, F., Peng, Z.: Diagnosis of OCD using functional connectome and riemann kernel PCA. In: Medical Imaging 2019: Computer-Aided Diagnosis, vol. 10950, pp. 610–620. SPIE (2019)
37. Xing, Y., Di Caterina, G., Soraghan, J.: A new spiking convolutional recurrent neural network (SCRNN) with applications to event-based hand gesture recognition. Front. Neurosci. **14**, 1143 (2020)

38. Yadav, S.P., Zaidi, S., Mishra, A., Yadav, V.: Survey on machine learning in speech emotion recognition and vision systems using a recurrent neural network (rnn). Arch. Comput. Meth. Eng. **29**(3), 1753–1770 (2022)
39. Zhang, L.: Building logistic spiking neuron models using analytical approach. IEEE Access **7**, 80443–80452 (2019)

DSNet: EEG-Based Spatial Convolutional Neural Network for Detecting Major Depressive Disorder

Min Xia[1], Yihan Wu[1], Daqing Guo[2], and Yangsong Zhang[1](✉)

[1] School of Computer Science and Technology, Laboratory for Brain Science and Medical Artificial Intelligence, Southwest University of Science and Technology, Mianyang 621010, China
zhangysacademy@gmail.com

[2] Clinical Hospital of Chengdu Brain Science Institute, MOE Key Lab for Neuroinformation, University of Electronic Science and Technology of China, Chengdu 610054, China

Abstract. Major depressive disorder (MDD) is a mental disease that has a severe negative impact on people's daily lives, which has become a leading global health burden. Previous neuroscience studies have proved that MDD patients have altered structural and functional connectivity between different brain regions compared to normal individuals. Measuring brain activities via electroencephalography (EEG) is a cost-effective and appropriate method for the detection of mental disorders such as depression. In addition, as deep learning (DL) is gaining attention in various research fields, increasing DL methods have been presented to diagnose depression. Inspired by these angles, this paper proposed an end-to-end spatial convolutional neural network (CNN) called DSNet for depression classification based on the resting-state EEG signals. Evaluated on a public dataset, our model obtained better classification performance with the accuracy of 91.69% via the leave-one-subject-out (LOSO) cross-validation strategy compared to other DL models. The experimental results demonstrate that DSNet can effectively extract information on spatial differences between depressed and normal individuals and could be a potential model for MDD detection.

Keywords: Major depressive disorder (MDD) · Electroencephalography (EEG) · Convolutional neural network (CNN)

1 Introduction

Major Depressive Disorder (MDD) is a mental illness characterized by persistent low mood, declining interest and the risk of suicide in severe cases [10], which could cause cognitive dysfunction and poor work productivity [6]. Therefore, early detection and treatment of depression can prevent patients from getting

X. Ying (Ed.): HBAI 2022, CCIS 1692, pp. 50–59, 2023.
https://doi.org/10.1007/978-981-19-8222-4_5

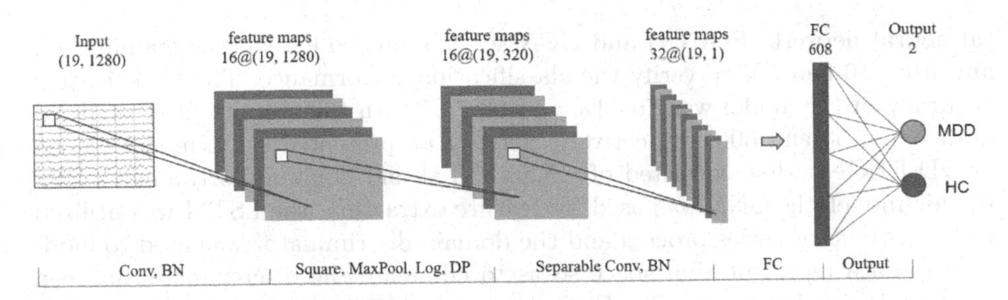

Fig. 1. The architecture of the proposed DSNet for MDD classification. The numbers above the picture represent the size of the feature maps after a series of specific operations indicated at the bottom of the diagram. Conv: the operation of convolution; DP: the operation of dropout.

worse. The current diagnosis of depression mainly relies on physician consultation and scale evaluation, for example, the Bech-Rafaelsen Mania Scale (BRMS) and the Hamilton Depression Scale (HDS) [3], but it is usually subjective and may lead to misdiagnosis. Compared to other physiological measurement tools, such as computed tomography (CT) [5] and functional magnetic resonance imaging (fMRI) [15], electroencephalography (EEG) has the advantages of high temporal resolution, low cost, ease of set-up and non-invasive technique [4], so it can be used as a quantitative measurement tool for the diagnosis of psychiatric disorders including depression.

Deep learning (DL) is a type of machine learning (ML) that automatically extracts representation from the input data. In recent years, it has been favored by researchers and widely used in various research fields, including motor imagery [2] and emotion recognition [24]. Accordingly, DL methods based on EEG signals have increasingly being applied to the detection of depression. Common DL algorithms include convolutional neural networks (CNN), long and short-term memory (LSTM), recurrent neural networks (RNN), graph convolutional neural network (GNN), a combination of CNN and LSTM, etc. For example, Zhang et al. conducted a 1-D CNN combined with an attention mechanism for the diagnosis of depression [25]. Demographic factors such as gender and age were added to the model by the attention mechanism to improve the classification accuracy. Based on the EEG data of 81 MDD patients and 89 healthy controls (HCs), the model achieved a classification accuracy of 75.29% with 10-fold cross-validation (CV). Seal et al. proposed a five-layer CNN model called DeprNet for the EEG signals of 15 depressed patients and 18 normal subjects classification [18]. The subject-wise split and the record-wise split CV strategies were used to validate the performance of the model, and classification accuracies of 99.37% and 91.4% were obtained respectively. Wang et al. proposed a semi-supervised GCN model for depression recognition [22]. The node and adjacency matrices of the graph were constructed by computing differential entropy features and their Pearson matrices based on the resting state EEG data of 24 depressed patients and 29 HCs. The model combined self-organizing incremental neural network (SOINN) and GCN self-training to extend the training set, and used 10-fold CV to verify the classification performance. The classification accuracy of the model was 70.53% and 92.23% with the amount of the labeled data being 50 and 600, respectively. Song et al. presented a DL model named LSDD-EEGNet that consisted of CNN, LSTM and domain discriminator [19]. In this model, the CNN was used for feature extraction, the LSTM was utilized to learn the time series process and the domain discriminator was used to modify the data representation space so as to eliminate the discrepancies between the training and test sets. The EEG data of 40 MDD patients and 40 HCs was used to validate the model performance by dividing the training and test sets at the subject level in a 7:3 ratio. The model eventually gained the classification accuracy of 94.69%.

Table 1. The detailed parameters of each layer of the DSNet.

No	Layer	Filters	Size	Activation	Output	Parameters
1	Input				$19 \times 1280 \times 1$	0
2	Conv2D	16	(1, 64)		$19 \times 1280 \times 16$	1024
3	BatchNormalization				$19 \times 1280 \times 16$	64
4	Activation			Square	$19 \times 1280 \times 16$	0
5	MaxPooling2D		(1, 4)		$19 \times 320 \times 16$	0
6	Activation			Log	$19 \times 320 \times 16$	0
7	Dropout(0.5)				$19 \times 320 \times 16$	0
8	SeparableConv2D	32	(1, 320)		$19 \times 1 \times 32$	5632
9	BatchNormalization				$19 \times 1 \times 32$	128
10	Activation			ReLU	$19 \times 1 \times 32$	0
11	Flatten				608	0
12	Dense			Softmax	2	1218

Moreover, several studies suggest that depressed patients have altered connectivity between brain regions compared to normal individuals and the findings suggest an increase in randomness of brain networks in MDD patients [14,16,23]. Inspired by these studies, the aim of this paper was to design an end-to-end CNN to diagnose depression by autonomously learning the spatial topological differences between MDD patients and HCs with a DL approach. Evaluated on a public dataset containing 30 MDD patients and 28 HCs, the proposed model yields better classification results than other comparison methods. The remaining part of this paper is organized as follows. The second part focuses on materials and methods, including data acquisition and processing, model architecture, baseline methods, model implementation and experimental evaluation. The third section presents the results of the experiment and discussion. The final section is a brief conclusion of this study.

2 Materials and Methods

2.1 Dataset and Data Preprocessing

To evaluate the classification performance of the proposed model and other comparison methods, a public dataset containing 30 MDD patients and 28 HCs was adopted in this study, which was provided by Mumtaz et al. at Hospital Universiti Sains Malaysia (HUSM) [17]. The experimental design was approved by the HUSM Ethics Committee. Five minutes of eyes-closed resting-state EEG signals were collected from each subject. Dataset was acquired by 19 electrodes following the 10–20 system [20]. The sampling rate of EEG signals was 256 Hz. The 50 Hz power line noise was eliminated via notch filter, and the interference signals were removed by band-pass filter of 0.5–70 Hz. Detailed information about the dataset refers to [17].

For the preprocessing of dataset, the EEG signals of each subject was first divided into multiple segments using a 5 s non-overlapping sliding window, i.e., each segment contained 5 * 256 = 1280 time points. Then, the segments with signal amplitudes out of [–100 uV, 100 uV] were excluded [13]. In total, 3167 samples were retained. Lastly, average reference and Z-score operations were performed for each sample. Moreover, we divided each sample into five frequency bands for further validation of the model, specifically the δ band (0.5–4 Hz), the θ band (4–8 Hz), the α band (8–13 Hz), the β band (13–30 Hz) and the γ band (30–70 Hz).

2.2 The Architecture of DSNet

In this study, we designed an end-to-end DL-based model named DSNet to automatic extraction of spatial topological differences between MDD patients and HCs. The structure of DSNet is depicted in Fig. 1. The model consisted of two convolutional layers, one pooling layer and one fully connected (FC) layer. The preprocessed EEG data was first input into the first convolutional layer with the kernel size of (1, 64) to learn features in the temporal dimension of each electrode. Next, the square activation function was utilized to capture the power information of the EEG signal. Then, the input data from the previous layer was down-sampled by a max-pooling layer with a pool size of (1, 4), which reduced the number of parameters and was followed by a log activation function to make the data smoother. Afterwards, the separable convolutional layer with a ReLU activation function extracted the spatial feature information between 19 electrodes through a convolutional kernel size of (1, N) (N indicated all sampling points on a single electrode). The separable convolutional layer essentially performed a two-step operation of depth-wise convolution and point-wise convolution, independently extracting the features of each input feature map and then optimally integrating these features for output, which lowered the model fitting parameters. Finally, the FC layer with a softmax activation function was deployed for binary classification. Each convolutional layer was immediately followed by a batch normalization (BN) layer to improve the training speed and accelerate the convergence process. The dropout layer can randomly cut out some neurons to prevent interactions between neurons that may cause unstable results and overfitting of the model. The detailed parameter settings of the DSNet are listed in Table 1.

2.3 Baseline Methods

In this section, two DL-based models are selected as baseline methods to compare the classification performance with our model. The general model architecture of the two baseline approaches is described below.

- EEGNet is a state-of-the-art (SOTA) model for EEG analysis [11], and some new models based on it were applied to different classification tasks [7,9,21]. It consisted of three different convolutional blocks, i.e. the temporal convolutional layer, the depth-wise convolutional layer and the separable convolutional layer. The first convolutional block utilized a large-scale convolutional kernel with the size of (1, 64) to learn the temporal properties of each channel. The depth-wise convolutional block with the kernel size of (19, 1) separately learned the spatial features of each feature map. The last block explored the relationships within and across feature maps through the separable convolutional layer with a small-scale kernel size of (1, 16). The first convolutional layer had no activation function while the following two convolutional layers used the ELU as an activation function. Each convolutional layer was followed by a BN layer.
- DeprNet is a DL model for depression classification [18], which had five convolutional blocks to discover features in the time dimension, and each convolutional layer was followed by a BN layer to stabilize the network and a max-pooling layer with a pool size of (1, 2) for the down-sampling of data. The number of filters used in the five convolutional layers was 128, 64, 64, 32 and 32 respectively. The filter size of the first three convolutional layers were (1, 5), and the filter size of the fourth and fifth convolutional layers were (1, 3) and (1, 2), respectively. The last three layers were FC layers with the number of neurons of 8, 4 and 2 respectively. The activation function of the last layer was softmax, and the activation function of the other layers was ReLU.

2.4 Model Implementation and Experimental Evaluation

In this study, the leave-one-subject-out (LOSO) CV strategy was employed to train and test the DSNet and baseline comparison models. The LOSO-CV method means that the training and testing data are divided on the subject level, with only one subject used for testing and the remaining subjects used for training in each round. The procedure was repeated 57 times (LOSO loops), such that each subject served as a test subject once. For each model, the training and testing operations were repeated 20 times in each LOSO loop [1], and the average value was taken as the final result of the corresponding testing subject.

Table 2. The classification performance of the DSNet and baseline models in the full band (mean ± std.).

Model	Accuracy(%)	Sensitivity(%)	Specificity(%)	Precision(%)	F1-score(%)
EEGNet	88.69 ± 1.44	88.12 ± 1.54	88.86 ± 2.29	88.98 ± 2.02	88.53 ± 1.28
DeprNet	88.54 ± 0.77	87.29 ± 1.65	88.51 ± 1.07	88.54 ± 0.88	87.90 ± 0.83
DSNet	$\mathbf{91.69 \pm 0.45}$	$\mathbf{92.11 \pm 0.58}$	$\mathbf{90.54 \pm 0.63}$	$\mathbf{90.82 \pm 0.56}$	$\mathbf{91.46 \pm 0.40}$

Table 3. The classification performance of three models in five frequency bands (mean ± std.).

	Band / Model	Delta	Theta	Alpha	Beta	Gamma
Accuracy(%)	EEGNet	86.02 ± 1.25	72.32 ± 1.64	**83.90 ± 0.83**	**88.87 ± 0.41**	85.52 ± 0.99
	DeprNet	86.40 ± 1.29	70.97 ± 1.10	74.13 ± 1.87	84.81 ± 0.68	86.03 ± 1.35
	DSNet	**88.59 ± 0.58**	**74.70 ± 0.74**	78.83 ± 0.80	88.09 ± 0.47	**90.03 ± 0.75**
Sensitivity(%)	EEGNet	**93.72 ± 0.84**	**85.12 ± 2.59**	**82.75 ± 0.91**	84.29 ± 0.71	86.42 ± 1.23
	DeprNet	87.32 ± 1.22	68.73 ± 1.62	71.95 ± 1.92	81.49 ± 1.48	86.44 ± 1.87
	DSNet	89.15 ± 0.62	73.80 ± 1.13	78.80 ± 1.28	**87.78 ± 0.50**	**92.38 ± 1.24**
Specificity(%)	EEGNet	75.77 ± 2.97	55.67 ± 2.42	**83.68 ± 1.22**	**93.00 ± 0.40**	84.83 ± 1.47
	DeprNet	83.95 ± 2.17	70.20 ± 1.27	74.67 ± 3.47	87.01 ± 1.05	85.33 ± 1.57
	DSNet	**86.40 ± 1.11**	**72.67 ± 1.31**	77.16 ± 0.98	87.15 ± 0.91	**86.86 ± 0.96**
Precision(%)	EEGNet	75.77 ± 2.97	55.67 ± 2.42	**83.68 ± 1.22**	**93.00 ± 0.40**	84.83 ± 1.47
	DeprNet	83.95 ± 2.17	70.20 ± 1.27	74.67 ± 3.47	87.01 ± 1.05	85.33 ± 1.57
	DSNet	**87.00 ± 0.92**	**73.31 ± 9.47**	77.82 ± 0.77	87.42 ± 0.78	**87.72 ± 0.77**
F1-score(%)	EEGNet	75.77 ± 2.97	55.67 ± 2.42	**83.68 ± 1.22**	**93.00 ± 0.40**	84.83 ± 1.47
	DeprNet	83.95 ± 2.17	70.20 ± 1.27	74.67 ± 3.47	87.01 ± 1.05	85.33 ± 1.57
	DSNet	**88.07 ± 0.57**	**73.55 ± 0.81**	78.31 ± 0.82	87.61 ± 0.44	**90.00 ± 0.07**

All experiments were implemented in Python with Keras framework. The binary cross-entropy loss was adopted as the loss function to calculate the cost of the model, and the Adam optimizer with a learning rate of 10^{-4} was utilized to update network weights. The network was trained with the batch size of 32 and the epochs of 30.

In addition, to evaluate the performance of the proposed framework for depression classification, four metrics, including sensitivity, specificity, precision and F1-score, were selected as the evaluation metrics.

Sensitivity is the probability that the model will correctly identify true positive samples, which is calculated as:

$$Sensitivity = \frac{TP}{TP + FN} \tag{1}$$

Specificity is the proportion that the model will correctly identify true negative samples, which is calculated as:

$$Specificity = \frac{TN}{TN + FP} \tag{2}$$

Precision indicates the proportion of true-positive samples out of predicted true-positive and false-positive samples, which is calculated as:

$$Precision = \frac{TP}{TP + FP} \tag{3}$$

F1-score can be considered as a summed average of the model sensitivity and precision, which is calculated as:

$$F1 - score = 2 * \frac{Precision * Sensitivity}{Precision + Sensitivity} \tag{4}$$

where TP and FN represent the number of MDD patients correctly and wrongly classified, respectively; TN and FP stand for the number of HCs correctly and wrongly classified, respectively.

3 Results and Discuss

We validated the classification performance of DSNet and two baseline models with the LOSO-CV method, whose classification results are presented in Table 2. It can be seen that DSNet yields better performance than EEGNet and DeprNet. The average classification accuracy is 91.69%, and the four evaluation metrics, including sensitivity, specificity, precision and F1-score are 92.11%, 90.54%, 90.82% and 91.46%, respectively. In addition, DSNet has the smallest standard deviation in each performance metric, thus indicating that our model is more stable over the 20 training sessions. The results demonstrate that DSNet could extract the hidden spatial topological features from the EEG signals.

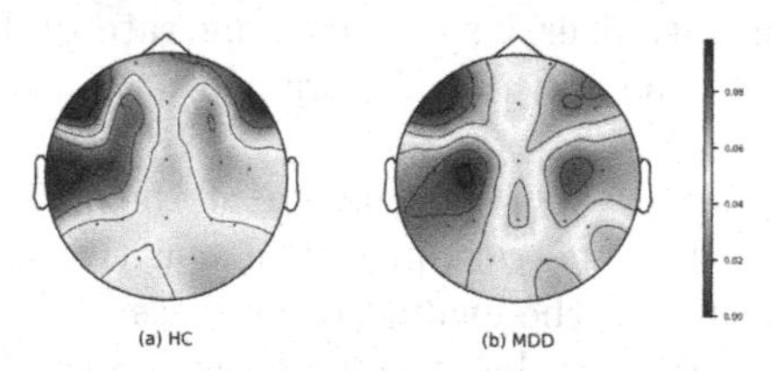

Fig. 2. The averaged topology represents the spatial features learned from the separable convolutional layer of all HCs subjects (left) and MDD subjects (right).

Furthermore, the EEG signals in five sub-bands, i.e., delta, theta, alpha, beta, and gamma were further utilized to calculate classification results with the three models, whose results are summarized in Table 3. The results reveal that DSNet outperforms two baseline models in the delta, theta and gamma band. In particular, the gamma band has the best classification performance compared with the other four sub-bands, which obtains the highest accuracy of 90.03% and is consistent with the findings in [19]. Thereby, it is suggested that the gamma band can be an essential biomarker in the detection of depression.

To verify the DSNet function on the mechanism by mining the spatial topological differences between MDD patients and HCs, we visualized the spatial features extracted by the separable convolutional layer. The output size of each subject from the 10th layer of DSNet was (samples, 19, 1, 32). The averaged topology is calculated as follows: First, we connected the output feature matrices of all MDD patients and HCs individually in the dimension of samples to obtain two fusion feature matrices. Second, Min-Max normalization was performed on the dimension of 19 electrodes. Finally, the averaged vector (19, 1) was calculated across samples and 32 filters, which was the averaged spatial topological features for the HCs (Fig. 2(a)) or MDD patients (Fig. 2(b)). By observing the topological map as shown in Fig. 2, it is found that HCs have higher values on the two sides of the prefrontal regions, while MDD patients have higher values on the two sides of the temporal regions, which proves that DSNet has captured the spatial difference information between the two groups. Besides, the results suggest that depression affects the activities of different brain regions, which is consistent with the findings of related neural mechanisms of depression [8,12].

4 Conclusion

This paper introduced an end-to-end model for the diagnosis of depression based on resting-state EEG signals. The DSNet was designed to exploit spatial topological differences in MDD patients and HCs. We conducted subject-independent experiments on the EEG signals of full-band and five sub-bands via the LOSO-CV strategy. The results demonstrated that the classification accuracy of DSNet in the full band exceeded that of the comparison baseline models by more than 3%. In addition, EEG signals in the gamma band could be better differentiated between MDD patients and HCs compared to the other four bands. The visualisation of the weights of the FC layer was eventually displayed for the interpretability of DSNet and the visualization results further explained the spatial discrepancy information automatically learned by DSNet. In the future, we will further evaluate the proposed method for other psychiatric diseases, such as schizophrenia, Alzheimer's disease, etc.

References

1. Acharya, U.R., Oh, S.L., Hagiwara, Y., Tan, J.H., Adeli, H., Subha, D.P.: Automated EEG-based screening of depression using deep convolutional neural network. Comput. Meth. Program. Biomed. **161**, 103–113 (2018). https://doi.org/10.1016/j.cmpb.2018.04.012

2. Al-Saegh, A., Dawwd, S.A., Abdul-Jabbar, J.M.: Deep learning for motor imagery EEG-based classification: a review. Biomed. Sign. Process. Control **63**, 102172 (2021). https://doi.org/10.1016/j.bspc.2020.102172
3. Bech, P., Bolwig, T., Kramp, P., Rafaelsen, O.: The bech-rafaelsen mania scale and the hamilton depression scale: evaluation of homogeneity and inter-observer reliability. Acta Psychiatrica Scandinavica **59**(4), 420–430 (1979). https://doi.org/10.1111/j.1600-0447.1979.tb04484.x
4. Biasiucci, A., Franceschiello, B., Murray, M.M.: Electroencephalography. Curr. Biol. **29**(3), R80–R85 (2019). https://doi.org/10.1016/j.cub.2018.11.052
5. Buzug, T.M.: Computed tomography. In: Springer Handbook of Medical Technology, pp. 311–342. Springer (2011). https://doi.org/10.1007/978-3-540-74658-4_16
6. Clark, M., DiBenedetti, D., Perez, V.: Cognitive dysfunction and work productivity in major depressive disorder. Expert Rev. Pharmacoecon. Outcomes Res. **16**(4), 455–463 (2016). https://doi.org/10.1080/14737167.2016.1195688
7. Deng, X., Zhang, B., Yu, N., Liu, K., Sun, K.: Advanced TSGL-EEGNet for motor imagery EEG-based brain-computer interfaces. IEEE Access **9**, 25118–25130 (2021). https://doi.org/10.1109/ACCESS.2021.3056088
8. Fingelkurts, A.A., Fingelkurts, A.A.: Altered structure of dynamic electroencephalogram oscillatory pattern in major depression. Biolog. Psychiatry **77**(12), 1050–1060 (2015). https://doi.org/10.1016/j.biopsych.2014.12.011
9. Huang, W., Xue, Y., Hu, L., Liuli, H.: S-EEGNet: electroencephalogram signal classification based on a separable convolution neural network with bilinear interpolation. IEEE Access **8**, 131636–131646 (2020). https://doi.org/10.1109/ACCESS.2020.3009665
10. Kennedy, S.H.: Core symptoms of major depressive disorder: relevance to diagnosis and treatment. Dialogues Clinic. Neurosci. (2022). https://doi.org/10.31887/DCNS.2008.10.3/shkennedy
11. Lawhern, V.J., Solon, A.J., Waytowich, N.R., Gordon, S.M., Hung, C.P., Lance, B.J.: Eegnet: a compact convolutional neural network for EEG-based brain-computer interfaces. J. Neural Eng. **15**(5), 056013 (2018). https://doi.org/10.1088/1741-2552/aace8c
12. Lee, T.W., Yu, Y.W.Y., Chen, M.C., Chen, T.J.: Cortical mechanisms of the symptomatology in major depressive disorder: a resting EEG study. J. Affect. Disord. **131**(1–3), 243–250 (2011). https://doi.org/10.1016/j.jad.2010.12.015
13. Lin, Y., et al.: Identifying refractory epilepsy without structural abnormalities by fusing the common spatial patterns of functional and effective eeg networks. IEEE Trans. Neural Syst. Rehabil. Eng. **29**, 708–717 (2021). https://doi.org/10.1109/TNSRE.2021.3071785
14. Liu, W., et al.: Functional connectivity of major depression disorder using ongoing EEG during music perception. Clinic. Neurophysiol. **131**(10), 2413–2422 (2020). https://doi.org/10.1016/j.clinph.2020.06.031
15. Logothetis, N.K., Pauls, J., Augath, M., Trinath, T., Oeltermann, A.: Neurophysiological investigation of the basis of the FMRI signal. Nature **412**(6843), 150–157 (2001). https://doi.org/10.1038/35084005
16. Mumtaz, W., Ali, S.S.A., Yasin, M.A.M., Malik, A.S.: A machine learning framework involving EEG-based functional connectivity to diagnose major depressive disorder (MDD). Med. Biologic. Eng. Comput. **56**(2), 233–246 (2018). https://doi.org/10.1007/s11517-017-1685-z
17. Mumtaz, W., Xia, L., Mohd Yasin, M.A., Azhar Ali, S.S., Malik, A.S.: A wavelet-based technique to predict treatment outcome for major depressive disorder. PloS one **12**(2), e0171409 (2017). https://doi.org/10.1371/journal.pone.0171409

18. Seal, A., Bajpai, R., Agnihotri, J., Yazidi, A., Herrera-Viedma, E., Krejcar, O.: Deprnet: a deep convolution neural network framework for detecting depression using EEG. IEEE Trans. Instrument. Measur. **70**, 1–13 (2021). https://doi.org/10.1109/TIM.2021.3053999
19. Song, X., Yan, D., Zhao, L., Yang, L.: LSDD-EEGNet: an efficient end-to-end framework for EEG-based depression detection. Biomed. Sign. Process. Control **75**, 103612 (2022). https://doi.org/10.1016/j.bspc.2022.103612
20. Teplan, M., et al.: Fundamentals of EEG measurement. Measure. Sci. Rev. **2**(2), 1–11 (2002)
21. Tsukahara, A., Anzai, Y., Tanaka, K., Uchikawa, Y.: A design of EEGNet-based inference processor for pattern recognition of EEG using FPGA. Electron. Commun. Japan **104**(1), 53–64 (2021). https://doi.org/10.1002/ecj.12280
22. Wang, D., et al.: Identification of depression with a semi-supervised GCN based on EEG data. In: 2021 IEEE International Conference on Bioinformatics and Biomedicine (BIBM), pp. 2338–2345. IEEE (2021). https://doi.org/10.1109/BIBM52615.2021.9669572
23. Zhang, B., Yan, G., Yang, Z., Su, Y., Wang, J., Lei, T.: Brain functional networks based on resting-state EEG data for major depressive disorder analysis and classification. IEEE Trans. Neural Syst. Rehabilit. Eng. **29**, 215–229 (2020). https://doi.org/10.1109/TNSRE.2020.3043426
24. Zhang, J., Yin, Z., Chen, P., Nichele, S.: Emotion recognition using multi-modal data and machine learning techniques: A tutorial and review. Inf. Fusion **59**, 103–126 (2020). https://doi.org/10.1016/j.inffus.2020.01.011
25. Zhang, X., Li, J., Hou, K., Hu, B., Shen, J., Pan, J.: EEG-based depression detection using convolutional neural network with demographic attention mechanism. In: 2020 42nd Annual International Conference of the IEEE Engineering in Medicine & Biology Society (EMBC), pp. 128–133. IEEE (2020). https://doi.org/10.1109/EMBC44109.2020.9175956

SE-1DCNN-LSTM: A Deep Learning Framework for EEG-Based Automatic Diagnosis of Major Depressive Disorder and Bipolar Disorder

Ziyu Zhao[1], Hui Shen[1](✉), Dewen Hu[1], and Kerang Zhang[2](✉)

[1] National University of Defense Technology, Changsha 410073, China
{shenhui,dwhu}@nudt.edu.cn
[2] First Hospital of Shanxi Medical University, Taiyuan 030012, China
atomsxmu@vip.163.com

Abstract. As two typical subtypes of depression, bipolar disorder (BD) is often misdiagnosed as major depressive disorder (MDD) in the early stage. Accurate diagnosis can provide effective treatment for patients. In this paper, we propose a deep learning framework namely SE-1DCNN-LSTM to automatically learn the latent EEG features of the two subtypes. Firstly, a SE block was used as a channel attention module to adaptively learn the weight of each electrode. Subsequently, a 1DCNN-LSTM network was applied to learn discriminative and effective patterns of EEG for MDD and BD. The noteworthy performance of proposed method was verified in 44 MDD and 26 BD patients with 81.10% and 83.16% classification accuracy in epoch-level and subject-level respectively. An extensive investigation of ablation analysis and window size of EEG epoch were conducted. Through visual analysis of electrode weights, we found that the weights of Fp1, Fp2, O1 and O2 electrodes were slightly greater. It demonstrated that the prefrontal lobe and occipital lobe may be possibly important brain regions for MDD and BD recognition. Overall, this study shows the effectiveness of the proposed model in EEG-based automatic diagnosis for MDD and BD.

Keywords: Major depressive disorder · Bipolar disorder · EEG · 1DCNN-LSTM · Channel attention · Deep learning

1 Introduction

Depression is one of the most urgent public health problems at present, and more than 350 million people around the world are troubled by it, which refers to a continuous terrible mental state even commit suicide in severe cases. Major depressive disorder (MDD) and bipolar disorder (BD) are two typical subtypes of depression. BD is a mental disease combined with both manic and depressive episodes, while MDD has single depressive episode. However, it is difficult to distinguish between MDD and BD in the early stage of depressive episode. Studies found that about 60% of BD patients had been misdiagnosed

X. Ying (Ed.): HBAI 2022, CCIS 1692, pp. 60–72, 2023.
https://doi.org/10.1007/978-981-19-8222-4_6

as MDD, and the average misdiagnosis time was 7.59.8 years [1, 2]. Furthermore, taking antidepressants in BD patients may increase the risk of manic attack, leading to far lower expected treatment effect [3]. Accurate diagnostic method of MDD and BD would allow effective treatment for patients, and reduce the deterioration of the disease.

Electroencephalogram (EEG) is a potential neurophysiological biomarker reflecting brain dynamics, featured with high time resolution, high sensitivity and nondestructive. As an objective method, EEG can reduce the interference of human subjective factors in the diagnosis process. In recent years, a large number of researchers have investigated the depression recognition based on EEG, and made some important progresses. Deep learning overcomes the shortcomings of traditional two-stage method consisted of manual feature extraction and classification without lots of prior knowledge to design and extract appropriate features, and thus it is more portable and friendly for clinical application. Moreover, the effective information learned in the end-to-end way may be richer. With the rapid development of deep learning in recent years, researchers try to apply deep learning to the automatic diagnosis of depression. [4] designed a 13-layer one dimensional convolutional neural network (1DCNN) for diagnosis of 15 MDD patients and 15 healthy subjects. The network used the input EEG signals of left and right hemispheres of subjects, adaptively learned and extracted feature for classification, and achieved the accuracy of 93.5% and 96% respectively. After that, [5] designed a network combined with 1DCNN and long short-term memory (LSTM) for the same 15 healthy subjects and MDD patients using the EEG signals in the left and right hemi-spheres, with an accuracy of 97.66% and 99.12%. [6] collected the resting EEG signals from 17 depression patients and 17 healthy subjects, expanded the samples through a sliding window, and compared the effects of 3s, 5s and 7s time windows. Finally, it obtained the best accuracy of 69.36% in epoch level and 73.53% in subject level respectively. [7] expanded the number of subjects to 33 MDD patients and 30 healthy subjects. Two different deep learning frameworks were designed to classify depression, namely 1DCNN network and 1DCNN- LSTM architecture, and obtained the accuracy rates of 98.32% and 95.57% respectively.

However, for the classification of MDD and BD based on EEG data, only two relevant literatures have been consulted. [8] based on the resting EEG signals of 48 BD patients and 55 MDD patients, utilized the improved ant colony optimization algorithm to select the coherent features of EEG in alpha, beta and theta bands, and then input into SVM classifier. 22 features were selected from 48 features for classification, and achieved 80.19% classification accuracy and 85.4% sensitivity. [9] utilized resting EEG signals of 31 BD and 58 MDD patients, extracted 28 coherent features in alpha and theta bands, and designed particle swarm optimization algorithm for feature selection. 14 features were selected for artificial neural network (ANN) classification, and the accuracy was 89.89%. Compared with [8, 9] used ANN as the classifier, and achieved higher accuracy, since the fitting degree of ANN to feature space may be better. However, both [8] and [9] are two-stage methods consisted of manual feature extraction and classification, requiring a lot of interdisciplinary prior knowledge, and the method of manual feature extraction may ignore some important information. In addition, neuroimaging studies on MDD and BD have shown that the functional connectivity between the posterior cingulate cortex and the right inferior parietal lobule, anterior central and insula in BD patients is significantly stronger than that in MDD [10]. Therefore, we hypothesized that the two subtypes of disorders are caused by abnormalities in specific brain regions, and thus the importance of EEG signals in different brain regions for diagnosis is different.

The purpose of this study is to explore the feasibility of deep learning in classification of MDD and BD based on EEG, and try to character the differences of abnormal brain regions related to MDD and BD. We propose a deep learning framework for automatic diagnosis of MDD and BD without any prior knowledge, and import the channel attention mechanism. To our knowledge, it is the first time to apply deep learning to the classification of MDD and BD based on EEG signals. Firstly, we input the preprocessed EEG signals into the channel attention module to obtain EEG signals with electrode weights, then input them into 1DCNN network to extract the timing characteristics, and the output of 1DCNN network is connected to the LSTM network. Finally, we use the full connected layer for the binary classification. The framework of this paper is organized as follows: the second part introduces the experimental materials and methods; In the third part, we compare the results with baseline method, and carry out ablation analysis. Then, the specific brain regions related to the two subtypes of disorder are analyzed based on the channel attention mechanism, and the effects of different time windows are discussed at last; The fourth part summarizes the work and puts forward the research prospect for the future work.

2 Materials and Methods

2.1 Data and Preprocessing

Subjects in the study were conducted in the outpatient department of First Hospital of Shanxi Medical University in China from 2015 to 2020. 70 patients' EEG data were collected including 44 MDD patients and 26 BD patients. The details of the subjects are shown in Table 1. No statistical difference exists in the age and gender distribution of the two subtypes. All participants signed written consent and were able to complete the whole experiment. The experimental procedure was implemented in accordance with the latest Helsinki declaration.

Table 1. Basic information of subjects.

	Age	Sex (M: F)	Number
MDD	13 ~ 49	18:26	44
BD	13 ~ 48	11:15	26
Total	13 ~ 49	29:41	70

EEG data of 2 min was collected for each subject with eyes-closed resting state. After eliminating the data segments with obvious noise, 90 s of EEG data of each subject was selected for subsequent analysis. The EEG data was acquired with 16 electrodes placed in accordance with the 10–20 electrode placement standard, and the average value of linked earlobe A1 and A2 was used for reference. The sampling frequency was 256 Hz. EEGLAB toolbox was used for essential preprocessing. Firstly, the EEG signals were filtered from 0.5–45 Hz by FIR filter with Hamming window. Subsequently, the independent component analysis (ICA) was applied to manually remove the noise of EMG and ECG and EOG.

Since the EEG signals of only 70 subjects were collected, the sliding window method for data augmentation was adopted to expand the samples for the 90s resting EEG signals of each subject [11]. We set the sliding window with 4s length for T, then shifted the window forward with 2s as the step S with 50% overlap, and finally obtained the ((90-T) / S + 1) segments for each subject.

2.2 1DCNN and LSTM Network

In this study, a SE-1DCNN-LSTM framework was proposed to extract the distinctive characteristics of two subtype subjects' resting EEG signals. The entire method framework was illustrated in Fig. 1, and the configuration of the model was summarized in Table 2. As illustrated in Fig. 1, the preprocessed epoch segments were firstly input into the 1DCNN network to extract the spatial and temporal features. Then, the output of 1DCNN network connected with the LSTM network to extract deeper temporal features. LSTM included input layer, hidden layer and output layer. Finally, the timing characteristics extracted by LSTM network connected to the two layers of fully-connected network for classification.

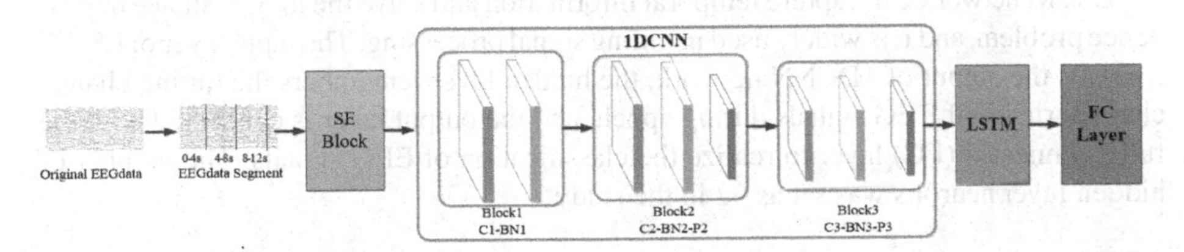

Fig. 1. The framework of the proposed SE-1DCNN-LSTM network.

The 1DCNN is a commonly used network in timing signal processing. The network structure of 1DCNN designed in this study is shown in Table 2. The network includes 3 blocks, and each block includes a convolution layer, a normalization layer or a pooling layer. The 1D convolutional layer uses 16 × 1 convolutional filter and 16 number of feature maps. Normalization layer was used to accelerate the convergence of the network and to reduce the overfitting of the network. Furthermore, the pooling layer can extract the large-scale features of the signal.

Table 2. Hyper-parameters of SE-1DCNN-LSTM

Block	Layer	Parameters	Output Shape
Input			(16,1024)
SE Block	Pooling		(16,1)
	FC1	(4)	(4,1)
	FC2	(16)	(16,1024)
1DCNN	Convolution(C1)	(1,16) (16)	(16,1009)
1DCNN	Convolution(C2)	(1,16) (16)	(16,994)
	Pooling(P2)	(1,4)	(16,248)
1DCNN	Convolution(C3)	(1,16) (16)	(16,233)
	Pooling(P3)	(1,4)	(16,58)
LSTM		32	(32,1)
FC	FC3	16	(16,1)
	FC4	2	(2,1)

LSTM network can capture temporal information and solve the long-distance dependence problem, and it is widely used in timing signal processing. The input layer of LSTM receives the output of 1DCNN network, the hidden layer remembers the timing change characteristics of EEG signals during epoch, and the output layer is connected with the fully connected (FC) layer to realize the classification of EEG signals. The number of hidden layer neurons was set as 32 in the study.

2.3 Channel Attention

The total 16 channels of EEG data were collected in the study, while different channels contributed differently to the recognition of MDD and BD. Figuring out the effective channels and brain areas for recognition of the two depression subtypes is essential for pathological and neuroscience study. Therefore, we designed a channel attention mechanism by Squeeze-and-Excitation (SE) block, which is a commonly used channel attention method in computer vision [12].

The structure of SE block is shown in Fig. 2. It consists of a global average pooling layer and two FC layers, while the activation functions are ReLU and Sigmoid functions respectively. The bottleneck structure with two fully-connected layers is adopted to reduce the complexity and improve the generalization ability of the model.

Firstly, the original EEG data with $C \times T$ size was input into the global pooling layer, where C represents the number of electrodes, and T denotes the length of signals. The global average of T data of each channel got a scalar, and thus obtained data with $C \times 1$ size, representing the average amplitude of EEG signals of each channel, which is called

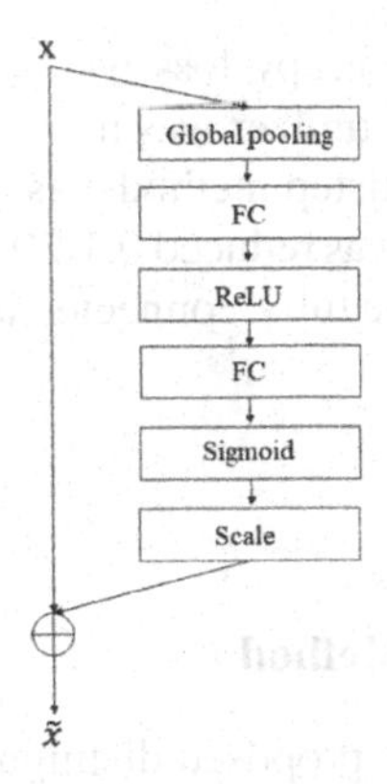

Fig. 2. The structure of SE block layer.

"Squeeze". Subsequently, the calculated average value of each channel was input into the two fully-connected network. Through the two fully-connected network, SE block can adaptively learn the weight of each channel and explicitly model the dependence among different channels. Finally, the obtained weights of all channels learned by the fully-connected network were applied to the original EEG signals to obtain new feature maps, which is called "Excitation". For each channel, a constant weight is predicted in this way and each channel is weighted.

In essence, the SE block performs an attention on the channel dimension, which allows the model to pay more attention to the channel features with the most relevant information and suppress those unimportant channel features. Therefore, the weights of channels with great importance for the binary subtype classification will increase in this way.

2.4 Evaluation Metrics and Parameters

In this experiment, A five-fold cross-validation was applied for evaluation. The classification of depression subtype is a typically non-subject-dependence task [13], so we divided the training set and test set according to 4:1 for 70 subjects, obtained the EEG signals of 56 subjects for training, and the remained 14 subjects for testing the model. As depicted in 2.1, the sliding window was used for expanding the samples, and 44 EEG data segments were obtained for each subject. Therefore, the 2464 training samples and 616 test samples were finally obtained for modeling.

Since the aim of the study is to classify MDD patients and BD patients, the recognition accuracy based on EEG epoch segments may not truly reflect the diagnosis in subject level. Hence, the classification accuracy for subjects of the model was also recorded. The proportion of all epoch segments predicted as a certain category of each subject was regarded as the probability of the subject diagnosed as the category, and the larger prediction probability class was determined as the prediction category in subject level. The classification accuracy (Acc), bipolar depression sensitivity (BD sensitivity), and area under curve (AUC) index were used as evaluation performance referred to [9].

In this experiment, the cross-entropy loss was determined as the loss function to train the network, and the Adam optimizer was implemented for all models. The initial learning rate was set to 0.001, multistep method was adopted to dynamically adjust the learning rate, and the learning rate was reduced 0.1/20 epoch. The batch size was 32, and the dropout to 0.5 was adopted in the fully-connected layer to reduce possible overfitting of the network.

3 Results and Discussion

3.1 Comparison with Baseline Method

To evaluate the effectiveness of the proposed distinguished method for MDD and BD, we compared the performance of our method with the commonly used baseline models, including graph convolutional neural network (GCNN) [14], EEGNET [15] and DeepConvNet [16]. For GCNN network, the phase-locked value (PLV) was extracted among channels to construct the graph adjacent matrix, and EEG signals were used as the node feature.

The loss value and accuracy of the networks during training process are shown in Fig. 3. It is obvious that all four networks converged during training, indicating the effectiveness of neural networks. As illustrated in Fig. 3, the SE-1DCNN-LSTM network converged the fastest and obtained the lowest loss. EEGNET network achieved the second-best performance, while GCNN got the relatively poor performance.

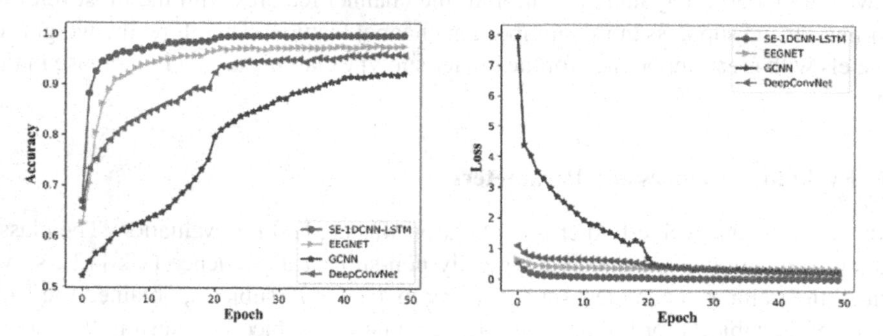

Fig. 3. Accuracy and loss of baseline networks during training process.

Table 3 displays the evaluation performance of models based on epoch-level and subject-level. Among all methods, the proposed SE-1DCNN-LSTM model obtained the highest classification accuracy both in epoch-level and subject-level, with the best accuracy of 81.10% and 83.16% respectively. The SE-1DCNN-LSTM network had an accuracy performance improvement of 3.26% and 2.87% compared to EEGNET, 13.41% and 7.21% compared to DeepConvNet, and 22.44% and 18.84% compared to GCNN in epoch-level and subject-level respectively. In addition, SE-1DCNN-LSTM obtained the best sensitivity of 84.9% in subject-level, indicating the effectivity of the proposed model for MDD and BD recognition. Furthermore, both the classification accuracy and BD sensitivity obtained comparable performance results with [8] using traditional two-staged method for binary classification of MDD and BD of 80.19% accuracy and 85.4% BD sensitivity. GCNN network got the relatively poor performance in this binary depression subtypes classification, and the reason may be the number of EEG electrodes was small and the graph adjacent relationship of each brain region may not be well expressed.

Table 3. Classification performance of baseline networks

Method	Epoch-Level(%)			Subject-Level(%)		
Performance	Acc	BD Sensitivity	AUC	Acc	BD Sensitivity	AUC
GCNN	58.66 ± 0.22	43.03 ± 1.64	54.48 ± 0.26	64.32 ± 0.36	40.00 ± 1.28	53.25 ± 0.68
EEGNET	77.84 ± 0.38	73.40 ± 2.96	75.76 ± 0.38	80.29 ± 0.88	78.67 ± 2.44	77.44 ± 0.96
DeepConvNet	67.69 ± 0.75	58.63 ± 1.13	65.93 ± 0.65	75.95 ± 1.71	76.57 ± 4.21	73.40 ± 1.97
SE-1DCNN-LSTM	**81.10 ± 0.16**	**76.74 ± 0.64**	**78.97 ± 0.24**	**83.16 ± 0.98**	**84.90 ± 2.20**	**80.45 ± 0.99**

3.2 Ablation Study

In order to further prove the effectiveness of our proposed method, the ablation experiments were implemented and analyzed. Ablation study was originally proposed by [17] to explore the importance of each module in neural networks.

We compare the performance of SE-1DCNN-LSTM network with 1DCNN, LSTM and 1DCNN-LSTM respectively. The training process and model performance based on epoch-level and subject-level are shown in Fig. 4 and Table 4.

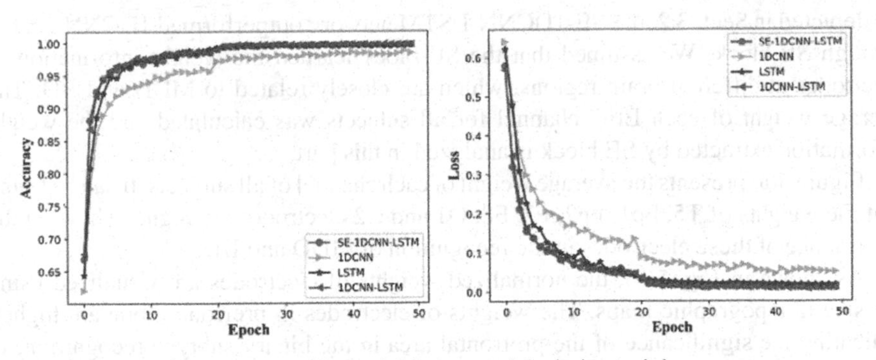

Fig. 4. Accuracy and loss of ablation models during training process

The 1DCNN-LSTM network and SE-1DCNN-LSTM network had the comparable performance during training process. All the 4 networks had good performance during training process, while 1DCNN network showed slightly poor performance. SE-1DCNN-LSTM had the highest classification accuracy both in in epoch-level and subject-level, which outperformed 1.38%, 1.22%, 0.54% than 1DCNN, LSTM and 1DCNN-LSTM of accuracy in epoch-level and 1.54%, 1.46%, 0.29% in subject-level. Furthermore, as for BD sensitivity in subject-level, SE-1DCNN-LSTM achieved 5.18%, 4.82% and 2.9% better performance than 1DCNN, LSTM and 1DCNN-LSTM, showing its strong sensitivity to MDD and BD patients.

Table 4. Classification performance of ablation models

Method	Epoch-Level(%)			Subject-Level(%)		
Performance	Acc	BD Sensitivity	AUC	Acc	BD Sensitivity	AUC
1DCNN	79.72 ± 0.87	72.81 ± 1.65	77.08 ± 0.82	81.62 ± 1.80	79.72 ± 3.60	79.68 ± 1.79
LSTM	79.88 ± 0.42	74.23 ± 0.72	77.37 ± 0.34	81.70 ± 0.65	80.08 ± 1.20	80.00 ± 0.77
1DCNN-LSTM	80.56 ± 0.50	76.20 ± 0.94	78.35 ± 0.48	82.87 ± 1.53	82.00 ± 1.91	80.23 ± 1.45
SE-1DCNN-LSTM	**81.10 ± 0.16**	**76.74 ± 0.64**	**78.97 ± 0.24**	**83.16 ± 0.98**	**84.90 ± 2.20**	**80.45 ± 0.99**

For 1DCNN network, the performance declined a little than 1DCNN-LSTM and SE-1DCNN-LSTM, and it may be the lack of LSTM layer for deeper temporal features. As for LSTM network, the accuracy dropped due to the lack of 1DCNN layer for proper temporal feature extraction. For 1DCNN-LSTM network, the lack of SE block led to slightly lower performance, which indicated the decisive importance of SE block.

Through SE block, EEG channels in the brain area related to the emotion of MDD and BD strengthened its weights respectively and played more important roles in subsequent recognition process.

3.3 Interpretability Analysis of Channel Attention

As depicted in Sect. 3.2, the SE-1DCNN-LSTM network outperformed 1DCNN-LSTM through SE block. We assumed that the SE block learned the weight information of electrodes at different brain regions, which are closely related to MDD and BD. The average weight of each EEG channel for all subjects was calculated and the weight information extracted by SE block is analyzed in this part.

Figure 5(a) presents the average weight of each channel of all subjects. It can be found that the weights of T5, Fp1, Fp2, F7, F8, O1 and O2 electrodes are higher, showing the importance of these electrodes in the recognition of MDD and BD.

As shown in Fig. 5(b), the normalized weights of electrodes are visualized using the spatial topographic maps. The weights of electrodes in prefrontal lobe are higher, indicating the significance of the prefrontal area in the binary subtype recognition. In the relevant research of neuroscience, it was found that the prefrontal lobe participated in the memory of events and waved emotion to the memory, and the prefrontal lobe was related to emotion and cognition [18]. In addition, the obtained result is consistent with [19], reporting that the elevated activity and volume loss of ventral prefrontal cortex and hypometabolism of dorsal prefrontal in MDD and BD patients. Moreover, [20] also reported that the inter-hemispheric frontal alpha asymmetry was identified as a distinct indicator in MDD patients. It can be inferred that some differences may exist in the prefrontal lobe of MDD and BD patients, and thus the electrodes in the prefrontal lobe are important for the binary subtype classification.

In addition, the weights of electrodes in occipital lobe were relatively high, since the occipital lobe may refer to emotion management, and [21] pointed out that the reduced activation of the occipital lobe may be an initiating factor for cognitive disorder in MDD patients. Therefore, there may also be differences in the occipital lobe of MDD and BD.

In summary, these results show that there are slight differences in weights of electrodes in the recognition of MDD and BD based on EEG data, and the weights of prefrontal lobe and occipital lobe are higher in the recognition. Furthermore, it can be inferred that the abnormal brain regions related to MDD and BD may be strongly correlated with the prefrontal lobe and occipital lobe, and there may be slight differences in the two brain areas of the two subtypes respectively, while further research may be carried out to prove this idea. This finding would make progresses in revealing the pathological mechanism of two disorders and would be helpful for accurate treatment.

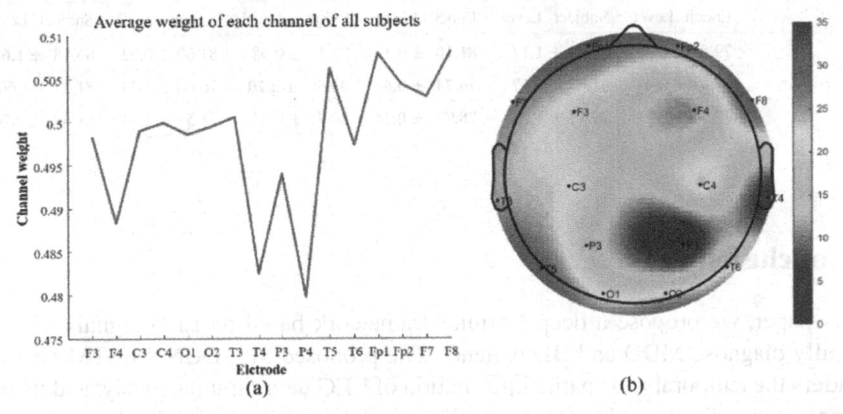

Fig. 5. **(a).** Average weight of each channel of all subjects. **(b).** Brain activation topographic map of average normalized weights of each channel.

3.4 Effects of Window Size

As for the window size of sliding window, large window may result in the attenuation of the number of expanded samples, which is not friendly to the deep learning algorithm involving huge samples for training, and the dimension of data may be high. If the window is too small, the information contained in each epoch may not be enough to reflect the stable difference of brain rhythmic activity between MDD and BD patients. Therefore, in this experiment, we investigate the effects of different window length with the same overlap for the binary classification and explore the sensitivity of the two depression subtypes to EEG signals with different windows. Specifically, two moving windows of 3 s and 5 s are selected for comparison, with the same 50% sliding overlap. For 90 s EEG data of each subject, 59 and 35 epoch EEG signals are generated respectively, and the corresponding 4130, 2450 samples were obtained for networks training.

The obtained results in Table 5 indicate that the classification performance of 4 s and 5 s window size are higher than that of 3 s window length, and it may be resulted by the less discriminative information contained in EEG signals with 3 s for recognizing MDD and BD. Therefore, the EEG epoch segment needs to be long enough to contain abundant information for disorder recognition. As a larger window size, the classification performance obtained by 5 s window size is quite equivalent to 4 s, while the BD sensitivity performance in subject-level of 4 s window size is slightly higher. Since the number of samples with 5 s window size is 20% lower than that with 4 s, the deep learning network is quite sensitive to the sample size, which may weaken the generalization ability of the network.

Table 5. Classification performance of SE-1DCNN-LSTM of different window size of EEG epoch

Performance (%)	Time window(s)					
	3 s		4 s		5 s	
	Epoch- Level	Subject- Level	Epoch- Level	Subject- Level	Epoch- Level	Subject- Level
Acc	79.54 ± 0.28	81.45 ± 1.17	**81.10 ± 0.16**	83.16 ± 0.98	81.00 ± 0.23	**83.18 ± 1.68**
BD Sensitivity	74.97 ± 0.41	81.07 ± 2.57	**76.74 ± 0.64**	**84.90 ± 2.20**	76.60 ± 1.24	81.54 ± 1.60
AUC	77.66 ± 0.34	79.11 ± 1.08	**78.97 ± 0.24**	80.45 ± 0.99	78.58 ± 1.54	**81.68 ± 1.70**

4 Conclusion

In this paper, we propose a deep learning framework based on EEG signals to automatically diagnose MDD and BD patients. The proposed SE-1DCNN-LSTM network considers the temporal and spatial information of EEG data simultaneously, and its performance outperforms with another baseline models and gets comparable results with traditional two-stage classification method, which shows the promise and good prospect of deep learning method for automatic diagnosis of MDD and BD. Since the clinical diagnosis is based on the subject level, we calculated the classification performance of epoch level and subject level. Through the visual analysis of the channel weights calculated by SE block, the interesting findings showed that the weights of Fp1 and Fp2 electrodes in prefrontal lobe and O1 and O2 electrodes in occipital lobe were greater in the recognition of the two subtypes, indicating that these areas may be strongly related to the binary classification, and there may be slight differences in the prefrontal lobe and occipital lobe for the two subtypes respectively. This finding may be helpful for the clinical diagnosis and analysis of pathological mechanism. In addition, we also discussed the effects of different time window size of EEG of 3 s, 4 s and 5 s for the classification, and suggested that the training EEG data epoch should be long enough for disorder recognition.

This manuscript shows the promise of deep learning framework in automatic diagnosis of MDD and BD. Future work will concentrate more on deep learning methods for automatic diagnosis of subtypes of mental health disease based on EEG data.

Acknowledgments. The authors would like to express thanks to First Hospital of Shanxi Medical University for providing the experimental EEG data.

References

1. He, H., Yu, Q., Du, Y., Victor, V., Victor, T.A., Drevets, W.C., et al.: Resting-state functional network connectivity in prefrontal regions differs between unmedicated patients with bipolar and major depressive disorders. J. Affect. Disord. **190**, 483–493 (2016). https://doi.org/10.1016/j.jad.2015.10.042
2. Hirschfeld, R., Cass A.R., Holt. D.C.L, Carlson.C.A.: Screening for bipolar disorder in patients treated for depression in a family medicine clinic. The J. American Board Family Medicine **18**(4), 233–239 (2005). https://doi.org/10.3122/jabfm.18.4.233
3. Ghaemi, S.N., Hsu, D.J., SoldaniF, F., Goodwin, F.K.: Antidepressants in bipolar disorder: the case for caution. Bipolar Disord. **5**(6), 421–433 (2015). https://doi.org/10.1046/j.1399-5618.2003.00074.x
4. Acharya, U.R., Oh, S.L., Hagiwara, Y., Tan, J.H., Adeli, H., Subha, D.P., et al.: Automated EEG-based screening of depression using deep convolutional neural network. Computer Methods Biomedicine Programs in Bio-medicine **161**, 103–113 (2018). https://doi.org/10.1016/j.cmpb.2018.04.012
5. Ay, B., et al.: Automated depression detection using deep representation and sequence learning with EEG signals. J. Med. Syst. **43**(7), 1–12 (2019). https://doi.org/10.1007/s10916-019-1345-y
6. Mao, W., Zhu, J., Li, X., Zhang, X., Sun, S.: Resting state EEG based depression recognition research using deep learning method. In: Wang, S., et al. (eds.) BI 2018. LNCS (LNAI), vol. 11309, pp. 329–338. Springer, Cham (2018). https://doi.org/10.1007/978-3-030-05587-5_31
7. Mumtaz, W., Qayyumb, A.: A deep learning framework for automatic diagnosis of unipolar depression. Int. J. Med. Informatics **132**, 103983 (2019). https://doi.org/10.1016/j.ijmedinf.2019.103983
8. Erguzel, T., Cumhur, T., Merve, C.: A wrapper-based approach for feature selection and classification of major depressive disorder–bipolar disorders. Comput. Biol. Med. **64**, 127–137 (2015). https://doi.org/10.1016/j.compbiomed.2015.06.021
9. Erguzel, T.T., Sayar, G.H., Tarhan, N.: Artificial intelligence approach to classify unipolar and bipolar depressive disorders. Neural Comput. Appl. **27**(6), 1607–1616 (2015). https://doi.org/10.1007/s00521-015-1959-z
10. Brooks, J.O., Wang, P.W, Ketter, T.A.: Functional brain imaging studies in bipolar disorder: focus on cerebral metabolism and blood flow. In: Yatham, L.N., Wang, P.W., Ketter, T.A.: (eds.) Bipolar Disorder. pp. 200–209. Wiley Online Library (2010). https://doi.org/10.1002/9780470661277.ch15
11. Lashgari, E., Liang, D., Maoz, U.: Data augmentation for deep-learning-based electroencephalography. J. Neurosci. Methods **346**, 108885 (2020). https://doi.org/10.1016/j.jneumeth.2020.108885
12. Hu, J., Shen, L., Albanie, S., Sun, G., Wu, E.H.: Squeeze-and-excitation networks. IEEE Transactions on Pattern Analysis and Intelligence Machine **42**(8), 2011–2023 (2020). https://doi.org/10.1109/TPAMI.2019.2913372
13. Fazli, S., Popescu, F., Danóczy, M.: Subject-independent mental state classification in single trials. Neural Netw. **22**(9), 1305–1312 (2009). https://doi.org/10.1016/j.neunet.2009.06.003
14. Song, T., Zheng, W., Song, P., Cui, Z.: EEG emotion recognition using dynamical graph convolutional neural networks. IEEE Trans. Affect. Comput. **11**(3), 532–541 (2020). https://doi.org/10.1109/taffc.2018.2817622

15. Lawhern, V.J., Solon, A.J., Waytowich, N.R., Gordon, S.M., Hung, C.P., Lance, B.J.: EEGNet: a compact convolutional network for EEG-based brain-computer interfaces. J. Neural Eng. **15**(5), 056013 (2016). https://doi.org/10.1088/1741-2552/aace8c
16. Schirrmeiste, R.T., Gemein, L., Eggensperger, K., Hutter, F., Ball, T.: Deep learning with convolutional neural networks for EEG decoding and visualization. Hum. Brain Mapp. **38**(11), 5391–5420 (2017). https://doi.org/10.1002/hbm.23730
17. Ren, S., He, K.M., Girshick, R., Sun, J.: Faster R-CNN: towards real-time object detection with region proposal networks. IEEE Trans. Pattern Analysis Intelligence Machine **39**(6), 1137–1149 (2017). https://doi.org/10.1109/TPAMI.2016.2577031
18. Gray, J.R., Braver, T.S., Raichle, M.E.: Integration of emotion and cognition in the lateral prefrontal cortex. Proc. Natl. Acad. Sci. U.S.A. **99**, 4115–4120 (2002). https://doi.org/10.1073/pnas.062381899
19. Hosokawa, T., Momose, T., Kasai, K.: Brain glucose metabolism difference between bipolar and unipolar mood disorders in depressed and euthymic states. Progress in Neuro-Psychopharmacology and Biological Psychiatry **33**(2), 243–250 (2009). https://doi.org/10.1016/j.pnpbp.2008.11.014
20. Kopecek, M., Barbora, T., Peter, S., Martin, B., Martin, B.: QEEG changes during switch rom depression to hypomania/mania: A case report. Neuro endocrinology letters. **29**(3), 295–302 (2008)
21. Li, J., Xu, C., Cao, X., Gao, Q., Wang, Y., Wang, Y.F., et al.: Abnormal activation of the occipital lobes during emotion picture processing in major depressive disorder patients. Neural Regen. Res. **8**(18), 1693–1701 (2013). https://doi.org/10.3969/j.issn.1673-5374.2013.18.007

Emotion Recognition from EEG Using All-Convolution Residual Neural Network

Hongyuan Xuan, Jing Liu, Penghui Yang, Guanghua Gu, and Dong Cui(✉)

Yanshan University, Qinhuangdao 066004, China
cuidong@ysu.edu.cn

Abstract. Emotion recognition has become a research hotspot due to the rapid development of machine learning and neuroscience. One of the most challenging tasks in the Brain Computer Interface (BCI) is to recognize human emotions by electroencephalography (EEG) signals. Motivated by the excellent performance of deep learning approaches in recognition tasks, we proposed an All-Convolution Residual Neural Network (ACRNN), which is a hybrid neural network that combines convolution neural network (CNN) and residual network (ResNet). The ACRNN solves the problem of information loss between convolution layer and full connection layer to some extent, and the time hardly increase. Meanwhile, instead of pooling layer, we increased the convolution step to reduce the size of the feature map, so there was no pooling layer in ACRNN. We conducted extensive experiments on the DEAP dataset to demonstrate the performance of the emotional recognition of the ACRNN. The experimental results demonstrate that the proposed method achieved an excellent performance with a recognition accuracy of 92.46% and 91.68% on arousal and valence classification task. It was verified that the ACRNN for emotion recognition is effective.

Keywords: Emotion recognition · ResNet · EEG · ACRNN

1 Introduction

With the rapid development of machine learning and neuroscience, interactions between the brain and devices, known as brain-computer interfaces (BCIs), have received great attention. Emotion recognition has always been an important research content of BCIs. Emotion plays an important role in human-computer interaction (HCI), and has been developed in various areas owing to its potential applications. Some disabled people with impaired voice or facial expression can express their emotions through fast and accurate emotion recognition. In addition, emotion recognition has been applied on depression [1] and autism spectrum disorder [2] to understand the emotional state of patients.

Researchers have made certain progress in feature extraction and data collection of emotion recognition and put forward a series of methods during the last decades. Emotion recognition methods can be divided into two categories. The first one is based on nonphysiological signals, such as facial expression and voice signal. The second one is based on physiological signal, such as electroencephalogram (EEG) [3] and electrocardiogram (ECG) [4]. Due to the physiological signals cannot be camouflaged and easy to

X. Ying (Ed.): HBAI 2022, CCIS 1692, pp. 73–85, 2023.
https://doi.org/10.1007/978-981-19-8222-4_7

collect, the researchers currently focus on the research of emotion classification based on physiological signals. Studying the brain mechanisms of emotion production and using machine learning to discriminate emotional states are the subject of several studies [5, 6]. For instance, several studies focus on emotion recognition from EEG signals. Some researchers study peripheral physiological data such as blood pressure [7], electrodermal activity signals [8]. Among these methods, the method based on EEG signals has the characteristic of directly reflecting emotional states, therefore, it has attracted a lot of attention from researchers. And in recent years, EEG signals have been widely used in the analysis of brain diseases [9, 10].

EEG signal-based emotion recognition methods usually consist of two parts, namely, EEG feature extraction and emotion classification. So far, researchers have conducted numerous studies on how to effectively extract features from EEG signals. The EEG features used for emotion recognition can be generally divided into three kinds, including time domain features, frequency domain features and time–frequency domain features. The time domain features capture the EEG emotion information from the temporal point of view, such as Frantzidis [12] et al. extracted the event related potentials (ERP) from EEG signals, and employed three different support vector machine (SVM) kernels for the emotional classification. Different from the time-domain feature, the frequency-domain feature aims to capture frequency information of EEG signals, such as Zheng et al. [13] extracted the power spectra density (PSD), differential asymmetry (DASM) and rational asymmetry (RASM) features of the 32-channel EEG data, and used SVM as a classifier. The methods of time-frequency feature extraction mainly include short-time Fourier transform, wavelet transform [14], etc.

In the research of EEG emotion classification algorithm, with the depth study of EEG signal classification, many researchers used deep learning approaches in emotion recognition tasks, because it has made a great success in pattern recognition domain. Wang [16] introduced three dimensional (3D) CNN to recognize the raw EEG signals, achieved classification rates of 73.3% and 72.1% for arousal and valence on DEAP dataset. Tang [11] used Dynamic Graph Convolutional Neural Network (DGCNN) to dynamically learn the internal relationship between different EEG channels, and achieved an average recognition accuracy of 90.4% for subject-related experiments. Since deep learning methods have shown excellent performance in emotion recognition based on EEG signals, we used deep learning methods in this paper.

In this paper, we proposed an All-Convolution Residual Neural Network model (ACRNN) for single modal emotion recognition based on EEG signals which can achieve higher accuracy. The remainder of this paper is organized as follows: Sect. 2 describes the method of data pre-processing, the all-convolutional neural network and deep residual learning framework for single-modal. Section 3 presents the contents of the experiment, including model implementation, experiment data and parameter settings. Section 4 gives the performance of the proposed model and shows the results of experiments. Section 5 presents conclusions and future works of the paper.

2 Methods

Considering the small amount of data in the EEG-based emotion classification task, the accuracy of the classification task cannot be improved by increasing the depth of the convolutional neural network, so the ResNet method is used to expand the width of the network and extract more features. Here we proposed an All-Convolution Residual Neural Network (ACRNN) that a hybrid neural network which combines convolution neural network (CNN) and residual network (ResNet). The ACRNN has an efficient representation of the spatial-temporal features of EEG signals when applied to recognize emotion states. The ACRNN solved the problem that the accuracy of training remains unchanged or even decreases with the increase of network layers, while the parameters and computational complexity remain unchanged.

2.1 Pre-processing and Feature Extraction

Pre-processing based on baseline signals is an effective way to improve the recognition accuracy in emotion recognition task [16]. The differences between signals which participants are under stimulation and baseline signals (under no stimulus) are calculated to represent the emotional state feature of this the segment. For the import multi-channel EEG data $X_{C \times H} \in R$, C is the number of channels and H is the data length, data segmentation of each channel is done first. We cut it in N segments with fixed length L, get N segments $C \times L$ matrixes. Then extract the baseline signals from segmented EEG data and calculate the value of Differential entropy (DE), next step calculated the value of DE from under stimulus EEG segmented and minus the value of DE from baseline signals, get the pre-processed signals.

DE for EEG-Based was used to measure the complexity of EEG signals and was proved effective [18], the principle of differential entropy algorithm is as follows.

DE can discretize the value of a continuous random variable, the differential entropy can be expressed as:

$$H(x) = -\int_x f(x) \log[f(x)]dx \tag{1}$$

$f(x)$ is the probability density function of signal x. For a signal x that obeying the Gauss distribution:

$$f(x) = \frac{1}{\sqrt{2\pi\sigma^2}} e^{-\frac{(x-\mu)^2}{2\sigma^2}} \tag{2}$$

the differential entropy can be expressed as:

$$H(x) = \frac{1}{2}\log\left(2\pi e\sigma^2\right) \tag{3}$$

2.2 3D Input Construction

3D EEG cubes were used as the input of the convolutional neural network, which can preserve spatial information among channels and combine features of signals from different frequency bands. In order to express EEG signals in different frequency patterns, wavelet transform (db4) was used to decompose the original signals into 4 frequency bands ($\theta(4-8\,\text{Hz})$, $\alpha(8-16\,\text{Hz})$, $\beta(16-32\,\text{Hz})$, $\gamma(32-45\,\text{Hz})$). Then the differential entropy values of all data segments in four frequency bands was calculated to represent the EEG features.

For preserving spatial information among multiple adjacent channels, the electrodes used were relocated to the 2D electrode topological structure. We transformed 1D DE feature vector to 2D plane ($h \times w$), where h and w is the maximum number of the horizontal and vertical used electrodes, this process is shown in Fig. 1. According to the EEG electrode map, here $h = w = 9$. The EEG electrodes filled with orange in EEG electrode map are the test points. The corresponding DE values are filled to construct a 2D equivalent matrix, and zeroes are used to fill the DEs at the un-test points.

Then the 3D EEG cubes with 4 frequency bands of 2D planes for each EEG segment are constructed to the inputs of the CNN.

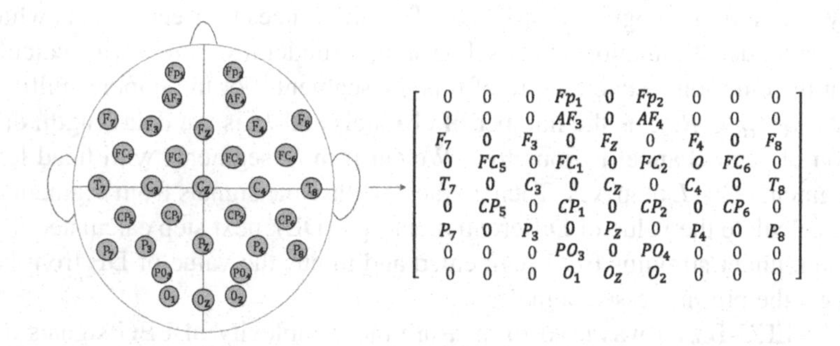

Fig. 1. The electrode distribution map.

2.3 The All-Convolutional Neural Network

We used an all-convolutional neural network (ACNN) with five convolutional layers to extract features from the input cube. As mentioned earlier, we transformed 1D DE feature vector to 2D plane, and constructed 3D EEG cubes with 4 frequency bands ($\theta, \alpha, \beta, \gamma$) of 2D planes. It is similar to those images are represented by Three Primary Colors (Red, Green and Blue) in computer vision. So, the constructed 3D EEG cube could be considered as color image with four primary colors ($\theta, \alpha, \beta, \gamma$), which allowed us to make full use of CNN as a powerful tool to extract representative features [19]. A fully connected layer with dropout operation was added for feature fusion. SoftMax layer was used for final classification. Instead of max-pooling, we used a convolution of stride, which was set to 2 in our model, so there was no pooling layer in the model. The process of ACNN is shown in Fig. 2.

The purpose of convolutional layer or pooling layer is different, but the feature map will be smaller after the convolutional layers or pooling layers. The purpose of convolutional layer is to extract features from the input, but the purpose of pooling layer is to reduce the dimension of feature by down-sampling. Down-sampling during pooling layer may filter out useful features, but convolutional layer controls the size of step and retains more features, thus we used an ACNN with five convolutional layers to extract features from the input. "all-convolutional" means that there are only convolutional

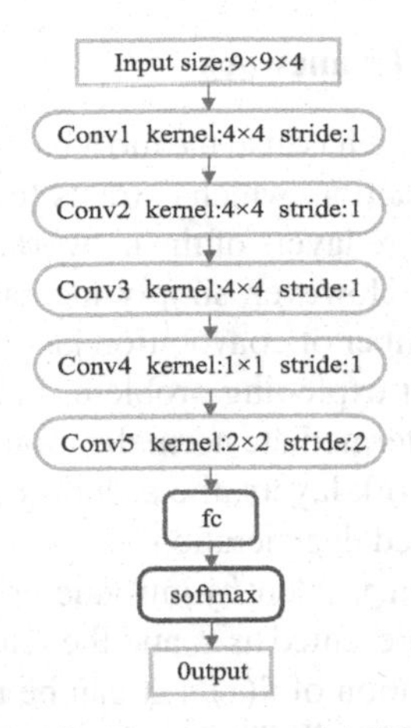

Fig. 2. The process of ACNN

layers and no pooling layer. Let f denote a feature map produced by some layer of a CNN, w and h are the width and height. Then p-norm subsampling (or pooling) with pooling size k (or half-length $k/2$) and stride r applied to the feature map f is a 3-dimensional array $s(f)$ with the following entries:

$$S_{i,j,u}(f) = \left(\sum_{h=-[k/2]}^{[k/2]} \sum_{w=-[k/2]}^{[k/2]} \left| f_{g(h,w,i,j,u)} \right|^p \right)^{1/p} \tag{4}$$

where $g(h, w, i, j, u) = (r \cdot i + h, r \cdot j + w, u)$ is the function mapping from positions in s to positions in f respecting the stride, p is the order of the p-norm (for $f \to \infty$, , it becomes the commonly used max pooling). If $r > k$, pooling regions do not overlap; however, current CNN architectures typically include overlapping pooling with k was set to 3 and r was set to 2. Let us compare the pooling operation defined by Eq. 1 to the standard definition of a convolutional layer c applied to feature map f given as:

$$c_{i,j,o}(f) = \sigma \left(\sum_{h=-[k/2]}^{[k/2]} \sum_{w=-[k/2]}^{[k/2]} \sum_{u=1}^{N} \theta_{h,w,u,o} \cdot f_{g(h,w,i,j,u)} \right) \tag{5}$$

where θ are the convolutional weights (or the kernel weights, or filters), $\sigma(\cdot)$ is the activation function, typically a rectified linear activation ReLU $\sigma(x) = \max(x, 0)$, and $o \in [1, M]$ is the number of output feature (or channel) of the convolutional layer.

According to the research made by Springenberg et al. [20], the pooling layer can be seen as performing a feature-wise convolution in which the activation function is replaced by the p-norm. The down-sampling of pooling layer can provide more receptive field for subsequent convolution operations. Avg-pooling is a general average filter convolution operation, while max-pooling introduces non-linearity and can be replaced by convolution layer with stride was set to 2. The performance is basically consistent or even slightly better.

2.4 Deep Residual Learning Framework

The input of CNN is a matrix, which is also the most basic feature. In fact, CNN network is a process of information extraction, which extracts features from the bottom layer to highly abstract features. The more layers of the network, the more abstract features can be extracted at different levels. However, simply increasing the depth of the network, that is to say, increasing the number of convolution layers, can easily cause the gradient vanishing problem and gradient exploding problem. In order to solve these problems, normalized initialization and intermediate normalization layers are introduced. Furthermore, with the increase of network layers, the accuracy of training remains unchanged or even decreases, which is called degeneration.

We addressed the degradation problem by introducing a deep residual learning framework. The input features are represented as x, and the features by one or more layers are represented as $f(x)$, the formulation of $f(x) + x$ can be realized by feedforward neural networks with "shortcut connections". Shortcut connections are those skipping one or more layers as shown in Fig. 3. Identity shortcut connections add neither extra parameter nor computational complexity [21]. In this paper, we used shortcut connections to optimize the all-convolutional neural network.

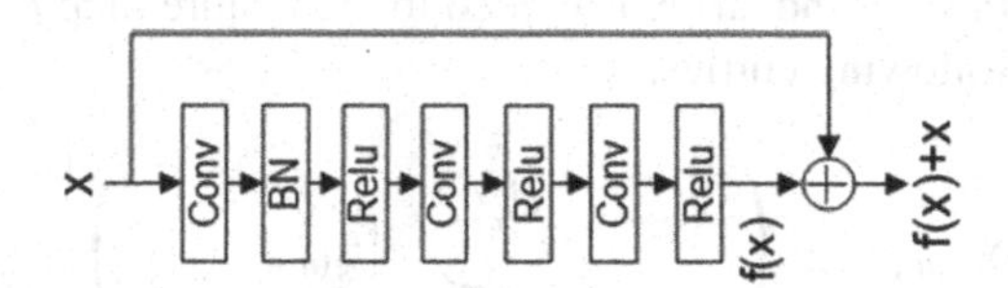

Fig. 3. Schematic diagram of shortcut connections.

3 Experiments

3.1 DEAP Dataset

The DEAP dataset is a large-scale dataset containing various physiological signals including EEG, which was widely used by various EEG-based emotion recognition researches. In the dataset, EEG signals of 32 participants (there are 16 male and female subjects, aged between 19–37 years old, with an average age of 26.9 years old) were recorded when they were watching 40 1-min-long videos. During the signal acquisition stage, EEG was recorded at a sampling rate of 512 Hz. Each video was presented to a subject and then she/he was asked to fill a self-assessment for her/his valence and arousal. Valence and Arousal scales are from 1 to 9 (1 represents sad/calm and 9 represents happy/excited) [22]. Based on the participant's evaluation values of each video, the evaluation values of valence and arousal are divided into two categories with the median of 5 as the threshold. The values more than 5 are marked as high Valence/Arousal and that less than or equal to 5 are marked as low Valence/Arousal. The 40 videos include 20 high price/arousal videos and 20 low price/arousal videos. The emotion distribution in Valence-Arousal space is shown in Fig. 4.

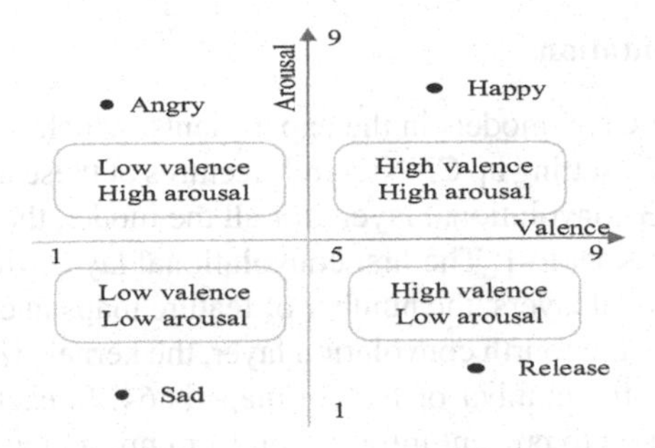

Fig. 4. Emotion distribution in Valence-Arousal space.

In this study, the EEG signals were down-sampled to 128 Hz and the EOG was removed. Furthermore, only the signal in the 0.5–45 Hz frequency bandwidth was preserved. The EEG signals with a total duration of 63-s (including 60-s for the participant to watch the video and 3-s before viewing) are shown in Table 1. In this paper, the EEG signals 3-s before watching the video was taken as the baseline, and the baseline signal was subtracted from the 60-s test signal to obtain the signal change related to the stimulus.

Table 1. The data used in DEAP dataset

Data	Numbers
Subjects	32
Videos	40
Channels	32
Sampling rate	128 Hz
Duration	63 s (60 s + 3 s)

The signals of four different frequency bands are extracted with db4 wavelet, which correspond to 4 frequency bands of EEG: $\theta(4-8\,\text{Hz})$, $\alpha(8-16\,\text{Hz})$, $\beta(16-32\,\text{Hz})$, $\gamma(32-45\,\text{Hz})$. The size of sliding window is $1\,s$. After decomposition, EEG data of a participant is converted from $40 \times 8064 \times 32(video \times sample \times channel)$ to $40 \times 8064 \times 4 \times 32(video \times sample \times band \times channel)$. After segmentation, EEG data is transformed into $40 \times N \times L \times 4 \times 32(video \times segment \times length \times band \times channel)$, , and is classified into correct labels, the differential entropy values (the differences between under stimulus EEG and baseline signals) of all data segments are calculated. We transform 1D DE feature vector to 2D plane $(h \times w)$, in DEAP dataset $h = w = 9$ Then 3D EEG cubes with 4 frequency bands of 2D planes for each EEG segment can be constructed, which are used as the inputs to the CNN.

3.2 Model Implementation

We used four different CNN models in the experiments, which were intended to reflect current best practices for setting up CNNs. Architectures of these networks are described in Table 2. In the first 3 convolutional layers for all the model, the kernel size was set to 4×4 and the stride was set to 1. The first convolutional layer with 64 feature maps. In the next two convolutional layers, the number of feature maps in each layer is twice that of the previous layer. In the fourth convolution layer, the kernel size was set to 1×1 and the stride was set to 1, the number of feature maps is 64. In each convolutional layer, zero-padding was applied to prevent information from missing at the edge of cube, BN layer and ReLU were also added after convolution. After these continuous convolutional layers, a fully connected layer was added to map the 64 feature maps (9×9) into a final feature vector $f \in R^{1024}$. Then the following SoftMax layer received f to predict human emotional state.

As described in the Table 2, the first four convolutions of the model have the same structure. In order to analyze the impact of reducing data dimensions on emotion classification results, we designed model B and model C based on model A. In model B, the max-pooling layer is added after four layers of convolution. We used an all-convolutional neural network as model C, 2×2 convolution in the fifth layer, and set the stride to 2. In order to verify the influence of residual module on emotion classification results, Model D is the ACRNN proposed in this paper, Shortcut connections were used to optimize all convolutional network and increase the width of the network. The network which after optimizing is shown in Fig. 5.

Table 2. The four different CNN models used for emotion recognition.

Model-A	Model-B	Model-C	Model-D
4×4conv stride = 1	4×4conv stride = 1	4×4conv stride = 1	4×4conv stride = 1
4×4conv stride = 1	4×4conv stride = 1	4×4conv stride = 1	4×4conv stride = 1
4×4conv stride = 1	4×4conv stride = 1	4×4conv stride = 1	4×4conv stride = 1
1×1conv stride = 1	1×1conv stride = 1	1×1conv stride = 1	1×1conv stride = 1
None	2×2 max-pooling. Stride = 2	2×2conv stride = 2	2×2conv stride = 2 add residual block
SoftMax	SoftMax	SoftMax	SoftMax

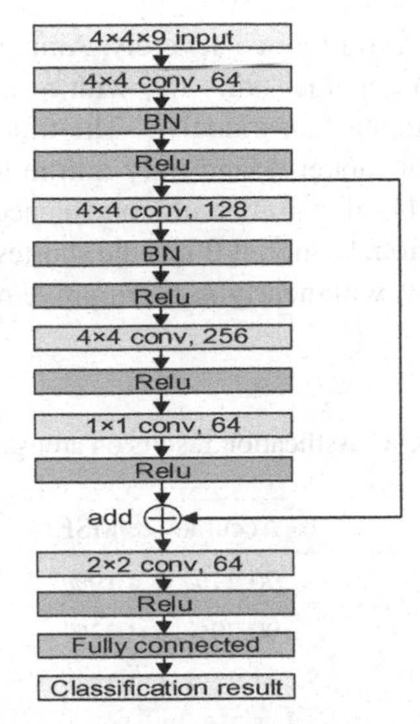

Fig. 5. All-Convolutional Residual Neural Network.

3.3 Parameter Setting

In this paper the time window length l was set to 128. The EEG data of a participant was cut into 60 segments, thus there were 2400 segments on each participant. In order to realize emotion recognition based on Arousal-Valence in two dimensions, we divided the labels into two binary classification problems.

The proposed model was implemented by using the Tensorflow framework. In the CNN models, the Adam optimizer was adopted to minimize the cross-entropy loss function and the truncated normal distribution function was used to initialize the weight of kernels. The initial learning rate was 10^{-4}. L2 regularization and Dropout was also adopted in CNN models, the penalty strength of L2 was 0.5 and the Dropout was set to 0.5. We used 20-fold cross validation on experimental data. The dataset was divided into 20 subsets, and 19 subsets were as train data and 1 subset as test data. The means of accuracy of each experiment were denoted as the results of the subject. In the 20-fold cross validation, because 19 samples were used to train the model, the distribution of the samples closest to the population and the estimated generalization error are more accurate. There are no random factors that affect the experimental data in the experimental process, so as to ensure the reproducibility of the experiment process.

4 Experimental Results and Discussion

To examine the impacts of model on the final classification results, we ran experiments on four models and compared their results. The same 3D EEG cube inputs were used to evaluate the classification accuracy of each model. The results for all the four models are shown in Table 3 and Table 4.

Tables 3 shows the classification results on arousal class. we can see that among four models, model B has the shortest time cost, and compared with model A, the classification accuracy is 90.79% which doesn't reduce severely. Among the four models, model D has the highest classification accuracy 92.46%, with nearly 2.2% improvement than model A and 1.2% improvement than model C. The time cost is 74934.2 *s*, , which is 11.36% shorter than that of model A and only a little longer than that of model B (2.42%). Table 4 shows the classification results on valence class, we can draw similar conclusions among the four models, model B has the shortest time cost, model D has the highest classification accuracy, with nearly 2.2% improvement than model A and 1.3% improvement than model C.

Table 3. Classification results on arousal class

Model	Accuracy ± MSE
A	90.27% ± 4.19%
B	90.79% ± 3.33%
C	91.22% ± 3.57%
D	92.46% ± 3.25%

Table 4. Classification results on valence class

Model	Accuracy ± MSE
A	89.45% ± 4.07%
B	89.99% ± 3.65%
C	90.39% ± 3.95%
D	91.68% ± 3.71%

Comparing the experimental results of models B and C, we can see that in the EEG-based emotion recognition task, it is feasible to use the method of increasing the convolution step size to replace the pooling layer, not only the training time is basically the same, but also more emotional features are retained to improve the classification accuracy. Comparing the experimental results of models C and D, it can be seen that in the EEG-based emotion recognition task, the added residual module expands the width of the network and improves the classification accuracy. This shows the effectiveness of Model D in the EEG-based emotion recognition task.

'We compared the proposed model with five other emotion recognition methods. All of these EEG emotion recognition methods were applied to the DEAP dataset. Xing et al. [23] extracted PSD from the raw EEG signals as a feature and used LSTM as a classifier for emotion recognition. Alhagry [15] et al. also employed LSTM but used raw EEG signals as inputs. Lin et al. [24] developed a 2D CNN model to conduct emotion recognition task using PSD. Salama et al. [25] applied a three-dimensional convolutional neural network (3D CNN) on raw EEG signals and achieved a state-of-the-art result with a mean accuracy of 87.97%. Yang et al. developed a parallel convolutional recurrent neural network to conduct emotion recognition and achieved mean accuracy of 90.92% with raw EEG signals. As shown in Fig. 6.

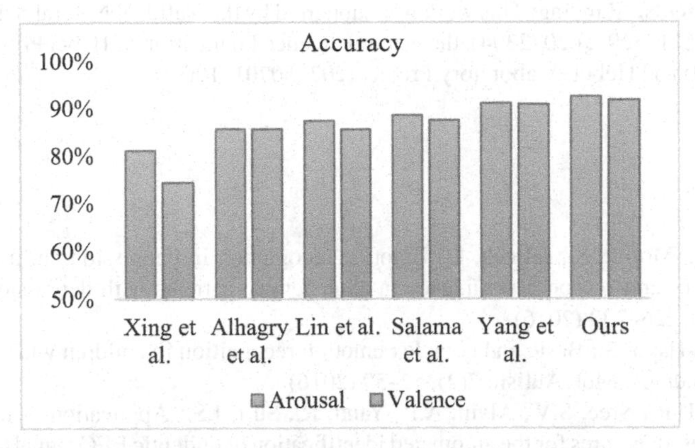

Fig. 6. Performance comparison between relevant approaches.

This paper summarizes the overall performance of our proposed model and other five different emotion recognition methods. The effectiveness of our method was observed in emotion classification for both arousal and valence. This result showed that our model had the best performance with an accuracy of 92.46% and 91.68% for arousal and valence. The mean accuracy achieved 92.07% of our mode, which is about 14% points higher than Xing et al., 6% points higher than Alhagry [15] et al. and 4.1% points higher than Salama et al. Lin and Yang et al. employed the similar model (2D CNNs) as our approach, achieved mean accuracy of up to 86.4% and 90.92%. This means that we have achieved better classification performance when using similar models. Otherwise, compared with original input EEG signals, DE features on our models are time consuming.

5 Conclusions and Future Works

Here we proposed an ACRNN model that consists of a 2D convolutional neural network model and a residual network model for EEG-Based emotion recognition. The residual network can directly connect the input to the deep convolutional layer. The biggest difference between the convolutional network and the residual network is that the deep network can directly learn the feature of the input. Traditional convolutional and fully connected layer may lose information when transmitting information, but Resnet solves this problem to some extent. The all-convolutional network after adding residual blocks improves the classification accuracy about 1.2% on Arousal-Valence (Table 3, 4), and the time cost will hardly increase.

We proposed an ACRNN model for EEG-Based emotion recognition is effective. Our results also confirmed that in the case of emotion recognition introducing residual blocks in CNN could add neither extra parameter nor computational complexity. Compared with other relevant CNN models, the proposed method has achieved the best performance with a mean accuracy of 92.46% and 91.68% for arousal and valence classification tasks on DEAP dataset.

Acknowledgments. Funding: This work was supported by the National Natural Science Foundation of China (62173291, 62072394), the Natural Science Foundation of Hebei Province of China (F2021203019) and Hebei ey aboratory Project (202250701010046).

References

1. Fieker, M., Moritz, S., Jelinek, L.: Emotion recognition in depression: an investigation of performance and response confidence in adult female patients with depression. Psychiatry Res. **24**(2), 226–232 (2016)
2. Fridenson-Hayo, S.: Basic and complex emotion recognition in children with autism: cross-cultural findings. Mol. Autism **7**(2), 52–59 (2016)
3. Acharya, U.R., Sree, S.V., Alvin, A.P., Yanti, R., Suri, J.S.: Application of non-linear and wavelet based features for the automated identification of epileptic EEG signals. Int. J. Neural Syst. **22**(2), 1250002–1250002 (2012)
4. Selvaraj, J., Murugappan, M., Wan, K., Yaacob, S.: Classification of emotional states from electrocardiogram signals: a non-linear approach based on Hurst. Biomed. Eng. Online **12**(1), 44–49 (2013)
5. Cowie, R.: Emotion recognition in human-computer interaction. IEEE Signal Process. Mag. **18**(1), 32–80 (2001)
6. Novak, M.J., Warren, J.D., Henley, S.M., Draganski, B., Tabrizi, S.J.: Altered brain mechanisms of emotion processing in pre-manifest Huntington's disease. Brain **135**(4), 1165–1179 (2012)
7. McCubbin, J.A.: Cardiovascular-emotional dampening: the relationship between blood pressure and recognition of emotion. Psychosom. Med. **73**(9), 743–750 (2011)
8. Al Machot, F., Elmachot, A., Ali, M., Al Machot, E., Kyamakya, K.: A deep-learning model for subject-independent human emotion recognition using electrodermal activity sensors. Sensors (Basel) **19**(7), 36–47 (2019)
9. Zhang, L., Chen, D., Chen, P., Li, W., Li, X.: Dual-CNN based multi-modal sleep scoring with temporal correlation driven fine-tuning. Neurocomputing **420**(2), 317–328 (2021)
10. Dong, H., Chen, D., Zhang, L., Ke, H., Li, X.: Subject sensitive EEG discrimination with fast reconstructable CNN driven by reinforcement learning: a case study of ASD evaluation. Neurocomputing **449**(4), 136–145 (2021)
11. Tang, Y., Chen, D., Li, X.: Dimensionality reduction methods for brain imaging data analysis. ACM Comput. Surv. **54**(4), 1–36 (2021)
12. Frantzidis, C.A., Bratsas, C., Papadelis, C.L., Konstantinidis, E., Pappas, C., Bamidis, P.D.: Toward emotion aware computing: an integrated approach using multichannel neurophysiological recordings and affective visual stimuli. IEEE Trans. Inf. Technol. Biomed. **14**(3), 589–597 (2010)
13. Zheng, W.-L., Zhu, J.-Y., Lu, B.-L.: Identifying stable patterns over time for emotion recognition from EEG. IEEE Trans. Affect. Comput. **10**(3), 417–429 (2019)
14. Mohammadi, Z., Frounchi, J., Amiri, M.: Wavelet-based emotion recognition system using EEG signal. Neural Comput. Appl. **28**(8), 1985–1990 (2016). https://doi.org/10.1007/s00521-015-2149-8
15. Alhagry, S., Fahmy, A.A., El-Khoribi, R.A.: Emotion recognition based on EEG using LSTM recurrent neural network. Int. J. Adv. Comput. Sci. Appl. **8**(10), 29–37 (2017)

16. Wang, Y., Huang, Z., McCane, B., Neo, P.: EmotioNet: a 3-D convolutional neural network for EEG-based emotion recognition. In: 2018 International Joint Conference on Neural Networks (IJCNN), pp. 1–7, Rio de Janeiro, Brazil (2018)
17. Kwon, Y.H., Shin, S.B., Kim, S.D.: Electroencephalography based fusion two-dimensional (2D)-convolution neural networks (CNN) model for emotion recognition system. Sensors (Basel) **18**(5), 1383–1395 (2018)
18. Duan, R.N., Zhu, J.Y., Lu, B.L.: Differential entropy feature for EEG-based emotion classification. In: 6th International IEEE/EMBS Conference on Neural Engineering, pp. 81–84, San Diego, CA, USA (2013)
19. Yang, Y., Wu, Q., Fu, Y., Chen, X.: Continuous convolutional neural network with 3D input for EEG-based emotion recognition. In: Cheng, L., Leung, A.C.S., Ozawa, S. (eds.) ICONIP 2018. LNCS, vol. 11307, pp. 433–443. Springer, Cham (2018). https://doi.org/10.1007/978-3-030-04239-4_39
20. Springenberg, J.T., Dosovitskiy, A., Brox, T., Riedmiller, M.: Striving for Simplicity: the All Convolutional Net (2014)
21. He, K., Zhang, X., Ren, S., Sun, J.: Deep residual learning for image recognition. In: 2016 IEEE Conference on Computer Vision and Pattern Recognition (CVPR) (2016)
22. Koelstra, S.: DEAP: a database for emotion analysis; using physiological signals. IEEE Trans. Affect. Comput. **3**(1), 18–31 (2012)
23. Xing, X., Li, Z., Xu, T., Shu, L., Hu, B., Xu, X.: SAE+LSTM: a new framework for emotion recognition from multi-channel EEG. Front. Neurorobot. **13**(1), 37–45 (2019)
24. Lin, W., Li, C., Sun, S.: Deep convolutional neural network for emotion recognition using EEG and peripheral physiological signal. In: Zhao, Y., Kong, X., Taubman, D. (eds.) ICIG 2017. LNCS, vol. 10667, pp. 385–394. Springer, Cham (2017). https://doi.org/10.1007/978-3-319-71589-6_33
25. Salama, E.S., El-Khoribi, R.A., Shoman, M.E., Shalaby, M.A.W.: EEG-based emotion recognition using 3D convolutional neural networks. Int. J. Adv. Comput. Sci. Appl. **9**(8), 329–336 (2018)

Salient Object Detection with Fusion of RGB Image and Eye Tracking Data

Yage Wu, Yadi Chen, and Rui Zhang(✉)

School of Electrical Engineering, Zhengzhou University, Zhengzhou 450001, China
ruizhang@zzu.edu.cn

Abstract. Salient object detection (SOD) is one of the fundamental topics in computer vision, but the current SOD algorithm is difficult to accurately find salient regions in scenarios such as multiple objects or small objects. Considering the above problems, this paper proposes a SOD algorithm with the fusion of RGB image and eye tracking data. The specific methods are as follows: (1) The eye tracking data can well simulate the human visual selection attention mechanism and contains high-level semantic information, so the eye fixation points are integrated into the salient object detection algorithm. (2) Considering the different characteristics of high-level features and low-level features, an improved cascade decoder including channel cascaded decoder (CCD) and spatial cascaded decoder (SCD) is designed. Moreover, the cross-modal fusion module (CMFM) is employed to better fuse the eye tracking data and the RGB image features. (3) The comparative experiments on the two datasets show that the performance of the proposed method exceeds that of the mainstream algorithms and can achieve effective SOD.

Keywords: Salient object detection · Eye tracking data · Cross-modal fusion module · Channel cascaded decoder · Spatial cascaded decoder

1 Introduction

Salient object detection (SOD) is identifying the areas or objects that are most interesting to people from a given scene. SOD can be used for pre-processing tasks in the computer vision, so it is widely utilized in image retrieval [1], image compression [2], image fusion [3], image recognition [4] and so on.

Most of the traditional SOD methods are based on some basic feature information of the image for heuristic search, such as color information [5], texture information [6], shape information [7] and so on., Combined prior knowledge, such as background prior [8], center prior [9], manually extract salient features through information comparison. Then, these features are used to calculate the saliency value of the image and obtain the saliency map. Traditional algorithms are devoted to the comparison of global and local information. Therefore, the

Supported by the Technology Project of Henan Province (no. 222102310031).

X. Ying (Ed.): HBAI 2022, CCIS 1692, pp. 86–98, 2023.
https://doi.org/10.1007/978-981-19-8222-4_8

algorithms are relatively easy to calculate, but these methods are often limited by the expressive ability of the low-level features and lack the advanced reasoning ability required for complex scenes.

In recent years, deep learning-based SOD algorithm has made great progress. Li et al. [10] first used deep neural networks to build a remarkable model based on multi-scale features. Hou et al. [11] proposed a deeply supervised salient object detection (DSSOD) model, which used a fully convolutional neural network to extract multi-level and multi-scale features and then fused the extracted multi-level and multi-scale features by introducing a skip-layer structure. Wu et al. [12] proposed a cascaded partial decoder (CPD) model, which integrated the deeper features in the backbone network to obtain the initial saliency map and then refined the features through the overall attention module to obtain the final saliency map. Liu et al. [13] proposed the pixel-wise contextual attention network (PiCANet) used by U-Net as the main backbone network to integrate global context information and multi-scale local context information, to further improve detection performance. BBS-Net [14] is a cascade optimization network using a bifurcated backbone strategy. First, a cascaded decoder is used to aggregate high-level features to generate initial saliency maps. Low-level features are refined by element-wise multiplication with the initial saliency map. Finally, the refined low-level features are integrated through another cascaded decoder to predict the final saliency map. Compared with the traditional salient object detection algorithms, the results are encouraging, but there are still some problems when the salient object and background have a similar appearance and the same object contains different colors.

We believe that the deep model-based SOD algorithm still faces two major challenges: (1) When salient object and background have a similar appearance, it is difficult to distinguish them effectively by RGB information only; (2) When the same object contains different colors, it is easy to be misjudged as different objects. However, eye tracking data can accurately record the position of human interest in the image. Therefore, the application of eye tracking data in SOD can better simulate the selection and attention mechanism of the human visual system, which is of great significance to image salient detection.

Eye tracking database records the movement information of human eyes when observing the image, and provides data support for understanding people's true intentions. Recently, eye tracking technology has attracted more and more attention and proved to have an important role in the significant detection of images. In the early methods, researchers predict the significance of the image by combining the characteristic information of different eye fixation points and images. Judd et al. [15] used the support vector machine to directly learn the visual model from the human eye tracking data and the characteristics of different levels. Xiao et al. [16] proposed a saliency detection model based on eye tracking data. The model first combined the eye tracking data to perform superpixel segmentation on the input image, extracted color and texture features for each superpixel, and then used the eye tracking data to select positive and negative samples and train the SVM classifier with the selected positive and negative samples, and finally classified each superpixel. Although eye tracking-based SOD methods have achieved excellent results, through experiments, it is found that the saliency map obtained by directly combining eye tracking data and low-level features is inaccurate, especially when the contrast between foreground and background is not obvious, the background cannot be well suppressed.

Inspired by the above-mentioned observations, this paper proposes a SOD algorithm with the fusion of RGB images and eye fixation points. The main contributions are as follows:

(1) To obtain more dimensions information about the images, the eye tracking saliency map (ETSM) is used to calibrate the fixation points of the input image, so as to estimate the relative position of the saliency region.
(2) In order to fully obtain useful information from ETSM and improve the compatibility of RGB and the characteristics of the ETSM, the cross-modal fusion module (CMFM) is established.
(3) To extract high-level features and low-level features more effectively, we designed the improved cascaded decoder with a channel cascaded decoder (CCD) and spatial cascaded decoder (SCD).
(4) A more effective optimization module (OM) is designed to obtain the sharpened object edge to effectively solve that the edge of the salient area of the complex image is relatively blurred.

2 Method

This paper uses ResNet-50 [17] as the backbone network for feature extraction and removes the last fully connected layer and pooling layer of ResNet-50. First, the ETSM is obtained through the eye fixation points, and then the ETSM is extracted through the five convolution blocks in the middle of ResNet-50. The feature map obtained is effectively fused with the feature C extracted from the RGB image through the CMFM. Get C_5. Divide the features of different levels into two groups, where $\{C_3, C_4, C\}$ is the high-level feature, and $\{C_1, C_2, C_3\}$ is the low-level feature. CCD aggregates high-level features to obtain initial saliency maps S_1. S_1 is multiplied with the low-level features to obtain the optimized low-level features $\{C_1', C_2', C_3'\}$. SCD aggregates to obtain the initial saliency map S_2, S_1 and S_2 finally go through the OM to obtain the final saliency map S. The overall framework of the proposed model is shown in Fig. 1.

2.1 Acquisition of ETSM

Both the DUT-OMRON dataset and the PASCAL-S dataset record the coordinates of the fixation points when observers observed the images on the monitor, and each image in the dataset records the eye tracking data of all the observers. The fixation points represent the location of people's interest, so the most interesting area in the image is the area with the densest fixation points distribution. First, we integrate the fixation points of all observers into a picture of the same

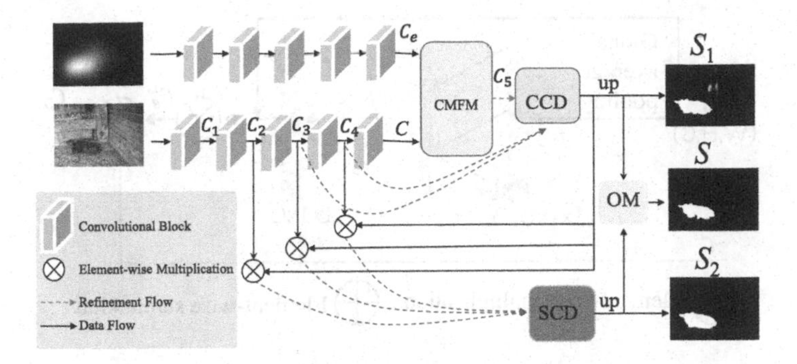

Fig. 1. Frame diagram of the model.

Fig. 2. Eye tracking saliency map.

size as the original image, and then calculate the distance influence relationship between each pixel on the picture and all the fixation points, as the eye tracking saliency value of the pixel, and then the S_{ETSM} is obtained, as shown in Fig. 2.

$$S_{ETSM}(m,n) = \sum_{i=1}^{j} e^{-\frac{(m_0(i)-m)^2+(n_0(i)-n)^2}{2\sigma^2}} \tag{1}$$

where $S_{ETSM}(m,n)$ represents the eye tracking saliency value at pixel (m,n), (m_0,n_0) is the coordinates of the fixation point. σ is the variance of the Gaussian function. In the experiment, we set σ^2 to 60. j is the number of eye fixation points in the image.

2.2 Cross-Modal Fusion Module

The ETSM contains a lot of fixation points, so it helps to locate the salient region, while the RGB image contains a lot of color and texture information. There is a modal difference between the two and cannot be treated in the same way. To fully obtain more useful information from the ETSM and improve the matching ability of RGB image and ETSM features, we add a CMFM. Inspired by [18], the matching of information fusion of different modalities is enhanced by using CMFM. The CMFM adopts a local cross-channel interaction strategy without dimensionality reduction, which greatly reduces the model parameters and improves the performance of the model. The input of CMFM is the feature C_e, extracted from the ETSM. First, the global features are aggregated through

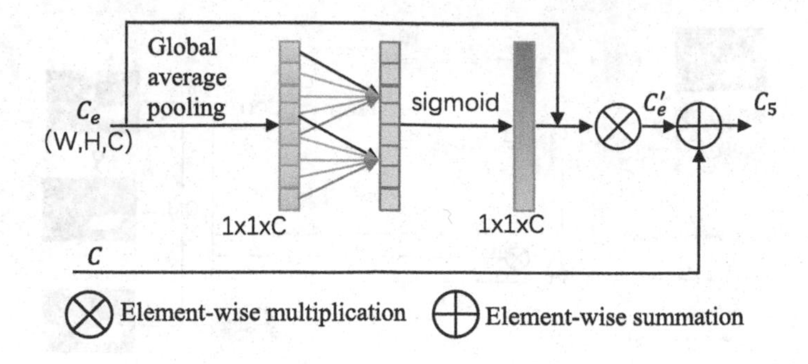

Fig. 3. Structure diagram of CMFM.

global average pooling, and then the channel weights are adjusted through the convolution operation. Then the weights are multiplied by the feature map C_e to obtain C'_e. Finally, C'_e and C performs element-wise addition to obtain the feature map C_5 after cross-modal fusion, and the structure diagram of CMFM is shown in Fig. 3.

2.3 Improved Cascade Decoder

Different levels of features are very different, and high-level features contain rich semantic information, which is conducive to locating region of interest location information. Low-level features are rich in detailed information and help to optimize edges. The mainstream salient object detection algorithms directly aggregate multi-level features without considering the characteristics of different levels of features and are easily disturbed by the background. The idea of [14] cascade optimization solves this problem, and the noise that may be caused by low-level features can be effectively suppressed by multiplying the initial saliency map with low-level features. Inspired by the idea of BBS-Net's cascade optimization, this paper proposes an improved cascade decoder including a CCD and SCD to aggregate high-level features and low-level features. The improved cascade decoder first considers that high-level features contain highly abstract semantic information, different channels may respond differently to foreground and background, and there is no need to filter spatial information. At the same time, there is almost no semantics between different channels in low-level features. It focuses on the importance of the features of a certain layer in different positions on the same plane. Compared with the traditional cascaded decoder [12], the CCD in the improved cascaded decoder adds a channel attention module (CAM) [19], which will pay more attention to the channel containing more key information, so it can obtain more abundant context information. The SCD in the improved cascade decoder adds a spatial attention module (SAM) [20], which filters out some background information.

The improved cascaded decoder includes the global contextual module (GCM), feature aggregation strategy module, CAM and SAM, as shown in Fig. 4. GCM is improved based on RFB [21], which can better capture the global

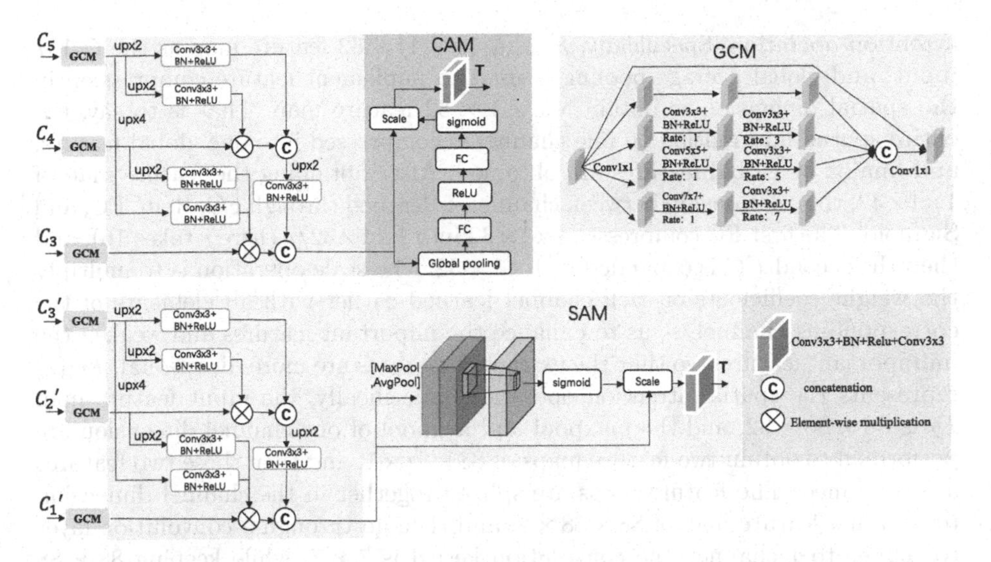

Fig. 4. Improved cascaded decoder.

environmental information. GCM consists of four parallel branches. Firstly, the number of channels is reduced through 1×1 convolution. For the second, third and fourth branches, the convolution of 3×3, 5×5 and 7×7 with an expansion rate of 1 is carried out respectively, and then the convolution operation of 3×3 with an expansion rate of 3, 5 and 7 is carried out respectively. After the convolution of 1×1, the number of channels becomes 32, and then it is connected with the residual of the input feature to get C_i^g.

The output of the global environment module for the improved cascade decoder is:

$$C_i^g = \begin{cases} F_{GCM}(C_i'), i \in \{1, 2, 3\}, (SCD) \\ F_{GCM}(C_i), i \in \{3, 4, 5\}, (CCD) \end{cases} \tag{2}$$

The input features of the aggregation strategy of the channel concatenated decoder and the spatial concatenated decoder are $\{C_3^g, C_4^g, C_5^g\}$, $\{C_1^g, C_2^g, C_3^g\}$, respectively. The specific calculation process is as follows:

$$C_i^{g'} = C_i^g \otimes \prod_{k=i+1}^{k_{max}} Conv(F_{UP}(C_k^g)) \tag{3}$$

where $k_{max} = \begin{cases} 3, i \in \{1, 2, 3\} \\ 5, i \in \{3, 4, 5\} \end{cases}$. $Conv(\cdot)$ is 3×3 convolution, F_{UP} represents up-sampling, $\otimes$ represents element-level multiplication. The output is then generated by a concatenation operation:

$$F = \left[C_k^{g'}; Conv\left(F_{UP}\left[C_{k+1}^{g'}; Conv\left(F_{UP}\left(C_{k+2}^{g'}\right)\right)\right]\right)\right] \tag{4}$$

Here, defining $F = \begin{cases} F_1, k = 3 \\ F_2, k = 1 \end{cases}$, $S_1 = T(F_{CAM}(F_1))$, $S_2 = T(F_{SAM}(F_2))$. T denotes two consecutive convolutional layers, and F_{CAM} represents the channel attention operation. Specifically, F_1 is a $44 \times 44 \times 32$ feature map as the CAM input, and global average pooling is used to implement feature compression in the spatial dimension to obtain a $1 \times 1 \times 32$ feature map. That is to say, the entire spatial information on one channel is compressed into one global feature, and finally, 32 global features are obtained. After obtaining the feature map of $1 \times 1 \times 32$, the relationship between channels is learned through FC, Relu, FC, and Sigmoid. The first FC compresses $1 \times 1 \times 32$ into $1 \times 1 \times 32/r$ (here r takes 16), and then the second FC is expanded to $1 \times 1 \times 32$. The scale operation is to multiply the weight coefficients of each channel learned earlier with all elements of the corresponding channel, so as to enhance the important features and weaken the unimportant features, so that the extracted features are more directional. F_{SAM} represents the spatial attention operation. Specifically, the input feature map F_2 is $88 \times 88 \times 32$, and the maxpool and avgpool of one channel dimension are performed to obtain two feature maps of $88 \times 88 \times 1$, and then these two features are combined. The feature maps are spliced together in the channel dimension to obtain a feature map of $88 \times 88 \times 2$, and then go through a convolution layer to reduce to 1 channel, the convolution kernel is 7×7, while keeping 88×88 unchanged, the output feature map is $88 \times 88 \times 1$, and then generated by the Sigmoid function The spatial weight coefficient is then multiplied with the input feature map to obtain the final feature map.

2.4 Optimization Module

In the improved cascaded decoder, the SCD is used to obtain accurate salient edge detail information, and the CCD is used to obtain contextual information with high-level semantics. Although the two kinds of information obtained by the above operation are complementary for saliency object detection. However, to better fuse high-level features and low-level features and obtain more accurate saliency maps, we designed an OM, the specific structure of OM in Fig. 5.

The input of the OM is S_1 and S_2. First, S_1 and S_2 are multiplied point by point, and the result obtained is added point by point with S_2. The expression is as follows:$S_3 = (S_2 \otimes S_1) \oplus S_2$.

To optimize the boundary blur and noise in the result, S_3 finally outputs the final saliency map S through the encode and decode layers. This part of the network is the classic encode-decode network. The previous encoding extracts features from the image and uses the down-sampling method to obtain high-level semantic features with gradually decreasing resolution. The latter part of the decoding is responsible for gradually restoring and enlarging the high-level semantic information. Thus, a feature map with a large resolution is gradually obtained, and the final output is a saliency map of the same size as the original image. Between encoding and decoding, there will be a shortcut to add feature maps of the same resolution, so that the final output feature map can take into account the features of high-level semantics and low-level semantics.

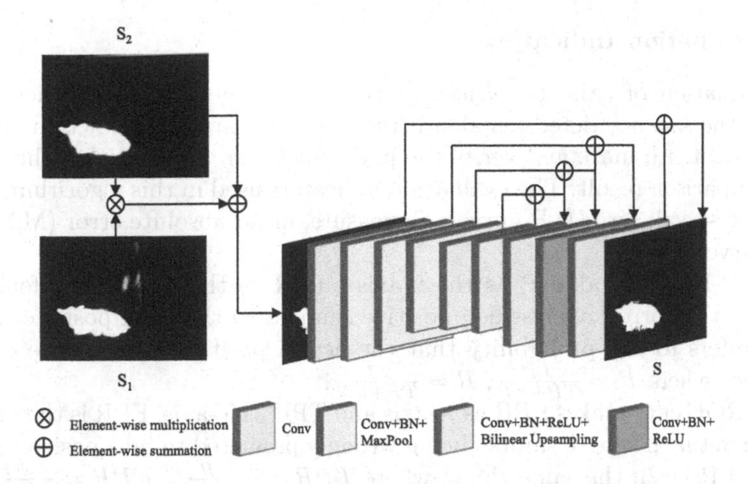

Fig. 5. OM structure diagram.

3 Experimental Setup and Result Analysis

3.1 Experimental Details

The training set is composed of 4000 images randomly sampled from the DUT-OMRON dataset and 400 images randomly sampled from the PASCAL-S dataset, and the test set is the remaining images of the two datasets. There are 5168 images in the DUT-OMRON dataset, and each image provides eye tracking data for 5 observers. All pictures are taken in natural scenes, and each picture scene is very complex, breaking the problem that the saliency algorithm is limited to a single dataset. The PASCAL-S dataset contains a total of 850 images, including eye tracking data from 8 observers.

The algorithm in this paper will compare the six currently popular salient object detection algorithms, and use their default training parameters to retrain the above models. The six methods are BASNet [22], poolnet [23], PSGLoss [24], CPD [12], EGNet [25], DACNet [26].

The model in this paper is implemented based on PyTorch and accelerated using an NVIDIA GeForce RTX 3060 GPU. All images are augmented with random flips, rotations, and border clipping during training. The initial learning rate in the experimental setting is 1e–4, and the optimizer is Adam.

The loss function used in this paper is binary cross entropy (BCE), and its calculation method is:

$$Loss = L_{BCE}(S, G) + L_{BCE}(S_1, G) + L_{BCE}(S_2, G) \tag{5}$$

where $L_{BCE}(N, G) = GlogN + (1-G)log(1-N)$, *Loss* is the total loss function, N is the predicted saliency map, and G is the ground-truth map.

3.2 Evaluation Indicators

The evaluation of saliency object detection is to generate a prediction map through the saliency detection algorithm, then compare the prediction map with the ground truth map, and verify the performance of the algorithm through the final comparison result. The evaluation indicators used in this algorithm are precision rate-recall rate (P-R curve), F-measure, mean absolute error (MAE), and ROC curve.

The P-R curve takes P as the x-axis and R as the y-axis. P refers to the proportion of correctly classified positive samples to the total positive samples, and R refers to the probability that the actual positive samples are correctly predicted, where $P = \frac{TP}{TP+FP}$, $R = \frac{TP}{TP+FN}$.

The ROC curve takes FPR as X-axis and TPR as Y-axis. FPR is the probability of an actual negative sample being wrongly predicted to be a positive sample. TPR and R mean the same thing, where $TPR = \frac{TP}{TP+FN}$, $FPR = \frac{FP}{FP+TN}$.

F-measure is the weighted harmonic average of recall and precision rate under non-negative weight β, where β^2 is set to 0.3. The formula is as follows: $F\text{-}measure = \frac{(1+\beta^2)P\times R}{\beta^2 P+R}$.

MAE is the difference between the predicted significance graph calculated and the real label, and formula is as follows: $MAE = \frac{1}{W\times H}\sum_{x=1}^{W}\sum_{y=1}^{H} |P(x,y) - G(x,y)|$.

3.3 Performance Comparison with Other Algorithms

Quantitative Analysis. To prove the effectiveness of the model in this paper, the mainstream saliency object detection methods, such as BASNet, CPD, poolnet, PSGloss, DACNet, and EGNet, were compared and tested, and the P-R curves and ROC curves (Figs. 6–7) of the corresponding datasets were obtained, and at the same time obtain a bar chart with MAE and F-measure as indicator values (Fig. 8). As shown in Figs. 6–7, in the P-R diagram, the method in this paper corresponds to the blue curve in the diagram. It can be observed that the P-R and ROC curves of the model can completely cover other curves, indicating that its performance is better than the other six SOD methods.

Compared with other similar models, the model in this paper has the best performance in the two indicators of MAE and F-measure in the two datasets, which quantitatively proves that the model in this paper has a good ability to detect salient regions and generate accurate saliency the ability to map.

Qualitative Analysis. To more intuitively show the superiority of the model in this paper, a visual comparison of different models is carried out, as shown in (Fig. 9). From the comparison, it can be seen that the model in this paper can not only highlight the object area, but also the scenes of small objects (line 8), low-contrast scenes (lines 2, 7), and multi-object scenes (lines 4, 6) can also effectively suppress background noise, while dealing with the problem of unbalanced

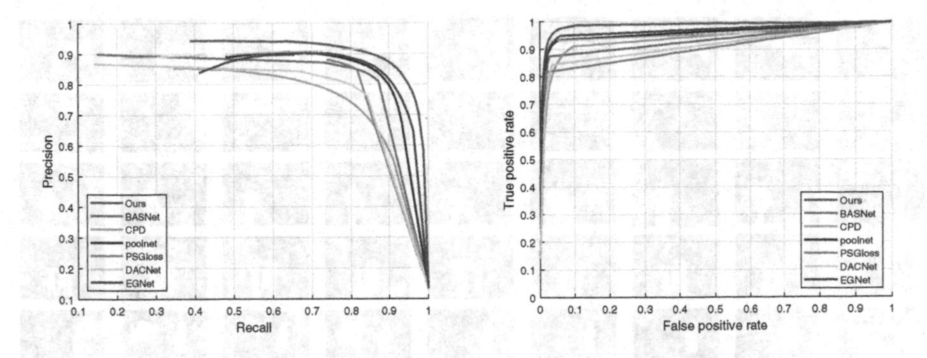

Fig. 6. P-R and ROC curves corresponding to the DUT-OMRON dataset.

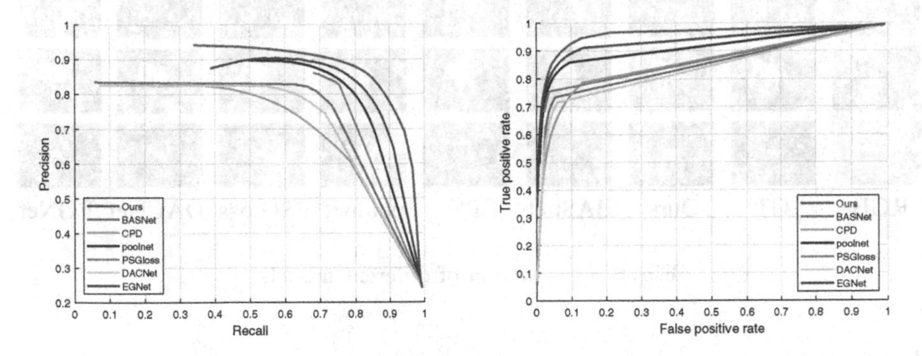

Fig. 7. P-R and ROC curves corresponding to the PASCAL-S dataset.

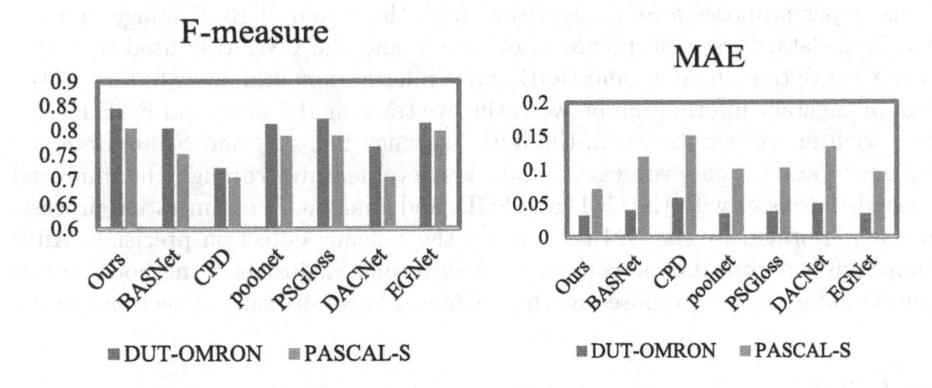

Fig. 8. F-measure and MAE indicators.

foreground and background. It can be seen from the images with poor segmentation effect (lines 1, 3, and 5) that the proposed model needs to be strengthened in processing detailed information, but its effect is obviously better than other models.

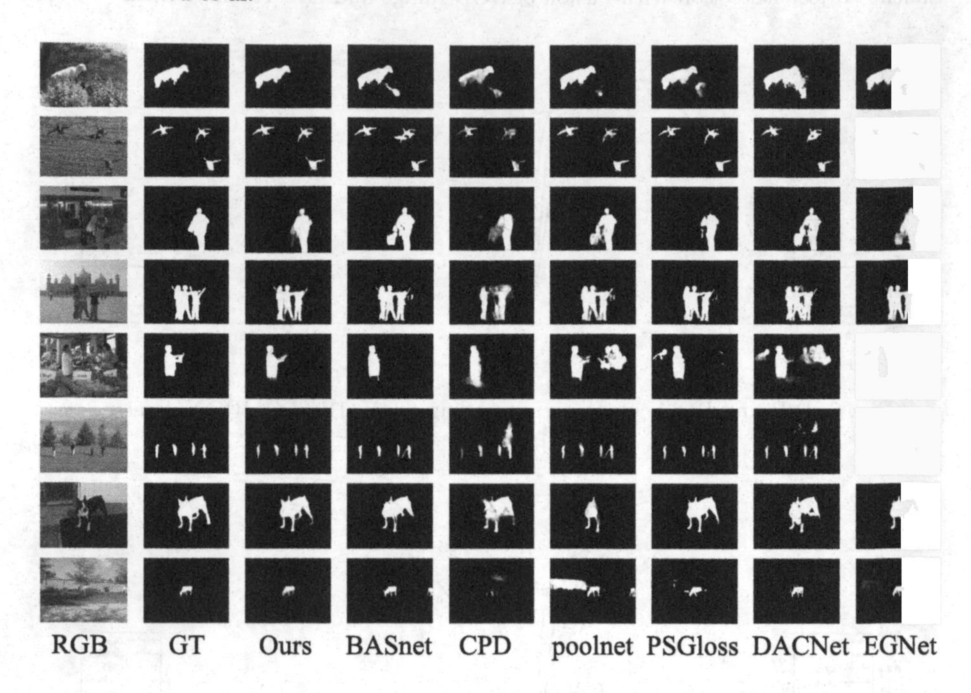

Fig. 9. Comparison of different models.

4 Conclusion

This paper proposes a SOD algorithm with the fusion of RGB image and eye tracking data. First, the ETSM is obtained, and the CMFM is used to better fuse the eye tracking data and RGB image information, and make full use of the complementary information between the eye tracking database and RGB images to enrich image details. Then, the initial saliency maps S_1 and S_2 are obtained by aggregating high-level features and low-level features through the improved cascaded decoder with the CCD and SCD, and finally edge optimization processing is performed by the OM to improve the saliency detection precision. After thorough experiments and comparisons with some mainstream methods on two public datasets, our proposed method achieves better saliency detection results.

References

1. Wu, H., Li, G., Luo, X.: Weighted attentional blocks for probabilistic object tracking. Vis. Comput. **30**(2), 229–243 (2014)
2. Shen, L., Liu, Z., Zhang, Z.: A novel h. 264 rate control algorithm with consideration of visual attention. Multimedia Tools Appl. **63**(3), 709–727 (2013)
3. Cheng, B., Jin, L., Li, G.: General fusion method for infrared and visual images via latent low-rank representation and local non-subsampled shearlet transform. Infrared Phys. Technol. **92**, 68–77 (2018)

4. Gao, D., Han, S., Vasconcelos, N.: Discriminant saliency, the detection of suspicious coincidences, and applications to visual recognition. IEEE Trans. Pattern Anal. Mach. Intell. **31**(6), 989–1005 (2009)
5. Cheng, M.M., Mitra, N.J., Huang, X., Torr, P.H., Hu, S.M.: Global contrast based salient region detection. IEEE Trans. Pattern Anal. Mach. Intell. **37**(3), 569–582 (2014)
6. Jiang, H., Wang, J., Yuan, Z., Wu, Y., Zheng, N., Li, S.: Salient object detection: a discriminative regional feature integration approach. In: Proceedings of the IEEE Conference on Computer Vision and Pattern Recognition, pp. 2083–2090 (2013)
7. Goferman, S., Zelnik-Manor, L., Tal, A.: Context-aware saliency detection. IEEE Trans. Pattern Anal. Mach. Intell. **34**(10), 1915–1926 (2011)
8. Zhu, W., Liang, S., Wei, Y., Sun, J.: Saliency optimization from robust background detection. In: Proceedings of the IEEE Conference on Computer Vision and Pattern Recognition, pp. 2814–2821 (2014)
9. Yang, C., Zhang, L., Lu, H., Ruan, X., Yang, M.H.: Saliency detection via graph-based manifold ranking. In: Proceedings of the IEEE Conference on Computer Vision and Pattern Recognition, pp. 3166–3173 (2013)
10. Li, G., Yu, Y.: Visual saliency based on multiscale deep features. In: Proceedings of the IEEE Conference on Computer Vision and Pattern Recognition, pp. 5455–5463 (2015)
11. Hou, Q., Cheng, M.M., Hu, X., Borji, A., Tu, Z., Torr, P.H.: Deeply supervised salient object detection with short connections. In: Proceedings of the IEEE Conference on Computer Vision and Pattern Recognition, pp. 3203–3212 (2017)
12. Wu, Z., Su, L., Huang, Q.: Cascaded partial decoder for fast and accurate salient object detection. In: Proceedings of the IEEE/CVF Conference on Computer Vision and Pattern Recognition, pp. 3907–3916 (2019)
13. Liu, N., Han, J., Yang, M.H.: Picanet: learning pixel-wise contextual attention for saliency detection. In: Proceedings of the IEEE Conference on Computer Vision and Pattern Recognition, pp. 3089–3098 (2018)
14. Fan, D.-P., Zhai, Y., Borji, A., Yang, J., Shao, L.: BBS-net: RGB-D salient object detection with a bifurcated backbone strategy network. In: Vedaldi, A., Bischof, H., Brox, T., Frahm, J.-M. (eds.) ECCV 2020. LNCS, vol. 12357, pp. 275–292. Springer, Cham (2020). https://doi.org/10.1007/978-3-030-58610-2_17
15. Judd, T., Ehinger, K., Durand, F., Torralba, A.: Learning to predict where humans look. In: 2009 IEEE 12th International Conference on Computer Vision, pp. 2106–2113. IEEE (2009)
16. Xiao, F., Peng, L., Fu, L., Gao, X.: Salient object detection based on eye tracking data. Sign. Process. **144**, 392–397 (2018)
17. He, K., Zhang, X., Ren, S., Sun, J.: Deep residual learning for image recognition. In: Proceedings of the IEEE Conference on Computer Vision and Pattern Recognition, pp. 770–778 (2016)
18. Wang, Q., Wu, B., Zhu, P., Li, P., Zuo, W., Hu, Q.: ECA-net: efficient channel attention for deep convolutional neural networks. In: 2020 IEEE/CVF Conference on Computer Vision and Pattern Recognition (CVPR), pp. 11531–11539 (2020)
19. Hu, J., Shen, L., Sun, G.: Squeeze-and-excitation networks. In: Proceedings of the IEEE Conference on Computer Vision and Pattern Recognition, pp. 7132–7141 (2018)
20. Woo, S., Park, J., Lee, J.Y., Kweon, I.S.: Cbam: convolutional block attention module. In: Proceedings of the European Conference on Computer Vision (ECCV), pp. 3–19 (2018)

21. Liu, S., Huang, D., et al.: Receptive field block net for accurate and fast object detection. In: Proceedings of the European Conference on Computer Vision (ECCV), pp. 385–400 (2018)
22. Qin, X., Zhang, Z., Huang, C., Gao, C., Dehghan, M., Jagersand, M.: Basnet: boundary-aware salient object detection. In: Proceedings of the IEEE/CVF Conference on Computer Vision and Pattern Recognition, pp. 7479–7489 (2019)
23. Liu, J.J., Hou, Q., Cheng, M.M., Feng, J., Jiang, J.: A simple pooling-based design for real-time salient object detection. In: Proceedings of the IEEE/CVF Conference on Computer Vision and Pattern Recognition, pp. 3917–3926 (2019)
24. Yang, S., Lin, W., Lin, G., Jiang, Q., Liu, Z.: Progressive self-guided loss for salient object detection. IEEE Trans. Image Process. **30**, 8426–8438 (2021)
25. Zhao, J.X., Liu, J.J., Fan, D.P., Cao, Y., Yang, J., Cheng, M.M.: EGNet: edge guidance network for salient object detection. In: Proceedings of the IEEE/CVF International Conference on Computer Vision, pp. 8779–8788 (2019)
26. Zhou, X., Fang, H., Liu, Z., Zheng, B., Sun, Y., Zhang, J., Yan, C.: Dense attention-guided cascaded network for salient object detection of strip steel surface defects. IEEE Trans. Instrument. Measure. **71**, 1–14 (2022)

Multi-source Domain Adaptation Based on Data Selector with Soft Actor-Critic

Qiquan Cui[1,2], Xuanyu Jin[1,2], Weichen Dai[1,2], and Wanzeng Kong[1,2(✉)]

[1] School of Computer Science, HangZhou DianZi University, Hangzhou, China
{cqqjerry,xy_jin,Weichendai,kongwanzeng}@hdu.edu.cn

[2] Key Laboratory of Brain Machine Collaborative Intelligence of Zhejiang Province, Hangzhou, China

Abstract. Multi-source domain adaptation (MDA) aims to transfer the knowledge learned from multiple-sources domains to the target domain. Although the source domains are related to the target domain, the difference of data distribution between source and target domains may lead to negative transfer. Therefore, selecting the high-quality source data is conducive to mitigate the problem. However, the existing methods select the data with uniform criteria, neglecting the variety of multiple source domains. In this paper, we propose a reinforced learning Data Selector with the Soft Actor-Critic (DSAC) algorithm for MDA. Specifically, the Soft Actor-Critic (SAC) algorithm has two Q-value Critic networks, it can better judge the performance of the data. Select the data in multi-source domains to migrate with our target domain, and use the difference in loss both before and after the model to determine the quality of the data and whether it is retained. Extensive experiments on the representative benchmark demonstrate that our method performs favorably against the state-of-the-art approaches.

Keywords: Multi-source domain adaptation · Reinforced learning data selector · Soft actor-critic

1 Introduction

Multi-source domain adaptation (MDA) addresses the adaptation from multiple source domains to a target domain. Although the source domain is related to the target domain, the difference in data distribution between the source and the target domain may lead to negative migration, and there are certain differences between multiple source domains. Choosing high-quality source data is conducive to reducing problems. However, the existing method has selected data with uniform standards, ignoring the diversity of multiple sources.

Learning a lot of resources in a large number of resources will suffer from a "Pyrrhic victory", where the quality of the data will significantly affect the performance of the model. Especially in natural language processing (NLP), this phenomenon is very important because noise and inaccurate comments are

X. Ying (Ed.): HBAI 2022, CCIS 1692, pp. 99–109, 2023.
https://doi.org/10.1007/978-981-19-8222-4_9

destroying the stability of the model in cross-domain applications [2,9]. Although various domain adaptation methods [11,21] are proposed for natural language processing tasks, most of them only consider scoring or ranking training data under a certain measurement of the entire data set, and then choose the top N (or proportional) project of the top N (or predetermined super parameter) project to be learned. However, this pre-designed indicator always enables efficient features of the transfer of domain knowledge or can be applied to different data properties. Despite the multi-function metrics, its super-parameter settings still need further exploration.

In this paper, we propose a reinforced learning Data Selector with the Soft Actor-Critic (DSAC) algorithm for MDA. Specifically, It combines the SAC algorithm, to encourage exploration, and it adds the concept of entropy. It unifies the problem of single source and multi-source domain adaptation, and the domain is good and bad by strengthening the rewards of learning. Strengthen the choice of learning to determine the domain and adjust the model by reward. Therefore, the alignment between the source domain and the target domain is significantly simplified because it no longer needs to align all the source domains with the target domain. In this case, we choose better samples to adapt to our model by strengthening learning methods to adapt to our model.

In [7,10,17], most works assume that static networks are hypothetical and focus on the loss function. The goal is to define the loss set of "align all domains" into a potential representation in some way. The problem is that domains are usually very different in-network input. This usually leads to difficult optimization and compromise adaptability. In this article, a SAC algorithm-based selector can be introduced, which can be more flexible mapping. In this case, there is no need to pull all the domains together. When the target domain is moved into the space formed by the entire source domain, the model selected by the learning selected can be adapted to the target domain through the feedback of the domain sample reward. In this way, the focus of the domain adaptation problem is shifted from the design of a good loss function design to good network architecture and a good sample for migration learning.

Based on this idea, we use the architectural design method explained in Fig. 1, and the two of them are the selectors and models. The selection is based on the SAC algorithm. It uses the selected sample for the migration of polygonal domains, and judges the quality of the source domain sample from the loss difference between the front and after the model as a reward. Our framework details are expanded in the following sections.

2 Related Work

Single Source Domain Adaptation: Single source domain adaptation methods modulate the model from the source to the target domain. The common method is to minimize the distance between the two domains. Although certain methods [10,17] minimize distance functions defined in the first and second-order data statistics, other methods can learn the potential space of cross-domain sharing by confrontation [16,18]. Although these methods are valid for single-source domain distribution and relatively simple data sets (for example, VisDA [14] or Office-31 [15]), due to more complex data, they do not have competitiveness for multi-source domain adaptation problems.

Multi-source Domain Adaptation: When the source contains a wide range of domains, consider the domain adaptation problem [23]. This problem is achieved by adaptive selecting the best choice in a set of assumptions for different source domains [1]. Based on $\mathcal{H} \triangle \mathcal{H}$ divergence, export the upper limit of the classification error that can be implemented in the target domain. Several methods have been proposed after introducing depth study. Some of these alignment domain pairs. Use the discriminator to align each source domain, and [13] matches all times all over the source domain and the target domain. These methods learn a classifier for each domain and use their weighted combinations to predict the category of the target sample [8]. Use mutual learning techniques to align the feature distribution of the source domain. Other methods focus on the critical alignment of all domains [20]. Model interacts with knowledge graphics Be2 Tween. Target sample prediction is based on their characteristics and relationships with different domains [5]. A meta-learning technology is proposed to search for optimal initial conditions of multi-source domain adaptive. Using an auxiliary network to predict the conversion ability of each source and use it as the weight of the learning domain discriminator. All these works select the data with uniform criteria, neglecting the variety of multiple source domains. In this paper, proposed a reinforced learning selector based on the SAC algorithm, select the source domain sample through the selector, and the model loss difference generated after training as a reward.

Data Selection: [19] proposed a method of selecting a model based on the Minimax Game for selective migration learning (MGTL), in this method, build a selector, identifier, and TL module to maximize the difference between samples. At the same time, this method studies the sample selection in a fixed threshold method. In this paper, we propose the method of strengthening the study selection sample based on SAC, which has a better judgment network to get better efficiency.

3 Method

In this section, we adapt to the enhanced learning selector for the multi-source domain, using the reinforce learning data selector from the source data set, transport to the shared encoder; strengthen the use of status, behavior, and rewards as screening mechanisms, screening the samples in the source domain.

3.1 Reinforcement Learning Data Selector

Use the reinforcement learning data selector to select the sample from the source data set, transfer it to the shared encoder; reinforce the learning data selector state, behavior, and rewards as the screening mechanism, and filter the sample in the source domain. The SAC algorithm used in the reinforcement learning data selector includes an Actor-network and four Critic networks.

Screening samples are input to the shared encoder as a training set; during training, the TL model and the reinforcing learning data selector jointly learn, further reserve, or delete the sample of the source domain. Target domain samples are identified using a well-trained TL model.

The status of the given source domain X_i is expressed as a continuous real value vector $S_i \in \mathbb{R}^l$, where l is the size of the status vector, and S_i represents the series:

- Hidden representation Z_i, it is the output of a given shared encoder;
- Training loss of source model X_i;
- Test loss of the target model Y_i;
- The prediction probability of shared encoder on the source model X_i;
- The prediction probability of shared encoder on the target model Y_i.

Behavior: It is represented as $a_i \in 0, 1$, which is used to delete or retain the sample from the source data, and sample a_i based on the probability distribution generated by the learning policy function $\Pi(S_i)$. The expression of $\Pi(S_i)$ is as follows:

$$\Pi(S_i) = softmax(W_2\ H_i + b_2) \tag{1}$$

$$H_i = \tan\left(W_l S_i + b_2\right) \tag{2}$$

where W_k and b_k are the weight matrix and bias carriers of the k layer in the policy network, $k = 1, 2, ..., l$; l is the number of layers in the policy network; Hi is the middle hidden state. Reward: The expression of the total reward $r\prime_b$ is as follows:

$$r\prime_b = \sum_{k=0}^{N-b} \gamma^k r_b + k \tag{3}$$

where N is the number of sample batches in this wheel; b is the current batch number; $r\prime_b$ is the expected total reward of the current batch b, γ is a reward discount factor; k is the policy network current layer serial number.

The update step of the SAC algorithm described in the selector is:

Target find the most flexible strategy π^*

$$\pi^* = arg\max_{\pi} \textstyle\sum_{t=0}^{T} \mathbb{E}_\pi[\gamma^t(R(s_t, a_t)) + \alpha\mathcal{H}(\pi(\cdot \mid s_t))] \tag{4}$$

where $\mathbb{E}_\pi$ is the expectation obtained under the policy π; $R\left(s_t, a_t\right)$ is a reward obtained by selecting behavior $a_t$in the s_t state; s_t is state; a_t is behavior t indicate the moment when α is the temperature coefficient, the importance of balancing the reward and strategic entropy given by the environment. $\mathcal{H}(\pi\Delta st)$ is entropy.

Build a flexible value $V^{\pi(s_t)}$ as follows:

$$V^\pi(s_t) = \mathbb{E}_{a_t \sim \pi(\cdot|s_t)}[Q(s_t, a_t) - \alpha log\pi(a_t \mid s_t)] \tag{5}$$

where $\pi(\Delta|st)$ is the probability of all actions, $Q\left(s_t, a_t\right)$ is actions function.

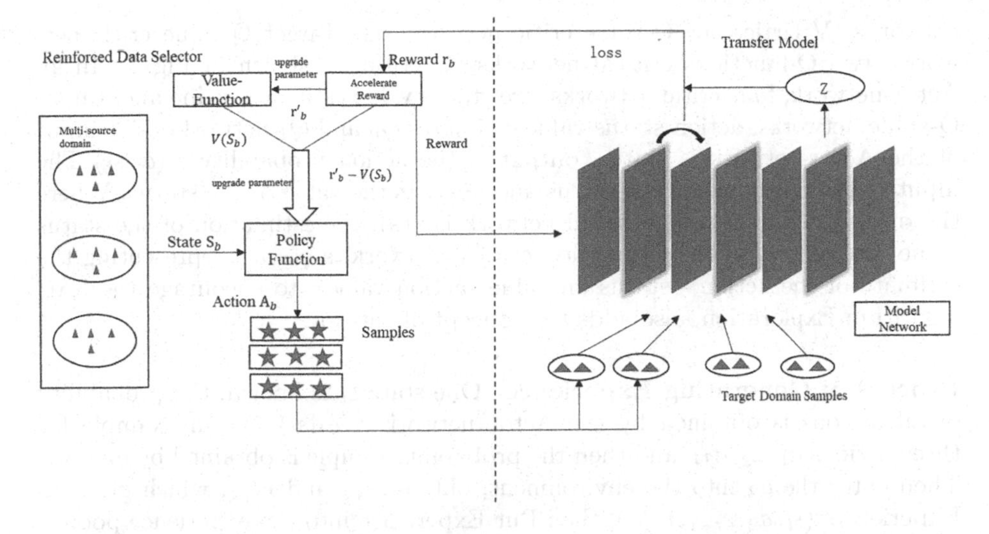

Fig. 1. Reinforcement learning data selector based on soft actor-critic.

Use the soft action value function $Q^{\pi}(s,a))$ in the environment in which the sample is learned is as follows:

$$Q^{\pi}(s,a) = R(s,a) + \gamma \sum_{s\prime \epsilon S} P(s'|s,a) V^{\pi}(s\prime) \tag{6}$$

where $V^{\pi}(s\prime)$ is the flexible value function under Status $s\prime$, $P(s'|s,a)$ the trajectory sample obtained from the environment, Build a flexible action value function $Q^{\pi}(s,a)$ as follows:

$$Q^{\pi}(s,a) = R(s,a) + \gamma \sum_{s\prime \in S} P(s'|s,a) Q^{\pi}(a', \pi(s')) \tag{7}$$

where $Q^{\pi}(a', \pi(s')$ is the flexible value function obtained in the previous policy.

Performing strategy is improved as follows:

$$\pi_{new} \longleftarrow arg \min_{\pi \epsilon \Pi} D_{KL}(\pi(\cdot \mid s_t)) \,||\, \frac{exp(\frac{1}{\alpha}Q^{old}(s_t,\cdot))}{Z^{\pi_{old}}(s_t)} \tag{8}$$

where π_{new} is the updated policy; $arg\min_{\pi\epsilon\Pi} D_{KL}(\pi(\cdot \mid s_t))$ is the smallest $\mathcal{D}_{KL}$; $\mathcal{D}_{KL}$ is the KL scattering of the experience pool (relative entropy), $Q^{old}(s_t, \Delta)$ for the Q value function under the previous policy, $Z^{\pi_{old}}(s_t)$ constant of gradient.

3.2 Soft Actor-Critic

This selector introduces Soft actor-critic, which is different from the ordinary actor-critic algorithm in that it is from one Actor-network, two Q-value critic networks (V cirtic, one Q-value critic network, one Target Q-value critic network), two Q-functions critical networks (Q critic, as shown in Fig. 2). In an Actor-network, four critic networks, are status value estimates value and Target Q-value networks, action-status value estimates Q_0 and Q_1 networks. The input of the Actor-network is status, output is the action probability $\pi(a_t|s_t)$ The input of the critic network is status and outputs the value of the status. Where the output of the Q-value critical network is $v(s)$, the estimation of the status value, the output of the Q-function critical network is $q(s, a)$, representing the estimate of the action - status on value (action value), to encourage the SAC algorithm Exploration, also adds the concept of entropy.

Process of Generating Experience. One state s_t is known, the probability of all actions is obtained by the Actor network $\pi(a_t|s_t)$ (as an example for three actions a_1, a_2, a_3), and then the probability sample is obtained by $a_t = a_2$, Then enter the a_2 into the environment, obtain s_{t+1} and r_{t+1}, which gives an Experience: $(s_t, a_2, s_{t+1}, r_{t+1})$, then Put Experience into the experience pool.

The meaning of the experience pool is to eliminate the correlation of Experience, because the reinforcing learning is usually strongly related, and dispels them, placed in the experience pool, and then in the training neural network, randomly from the experience pool a group of experience can make neural network training better Fig. 1.

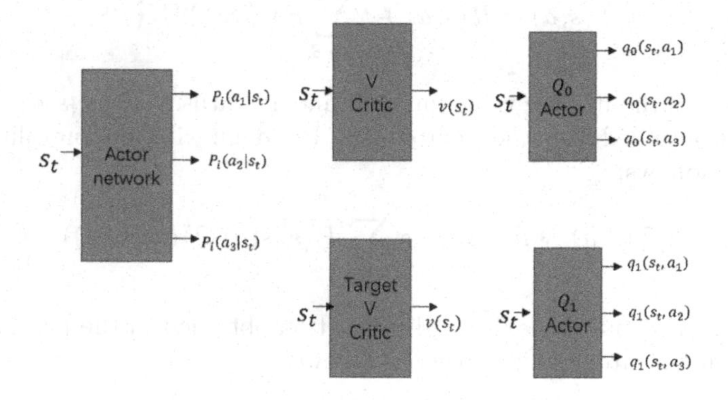

Fig. 2. Soft actor-critic.

Q-function Critical Network Update. Take the data from the experience pool buffer $(s_t, a_t, s_{t+1}, r_{t+1})$ for the critic network update, $(s_t, a_2, s_{t+1}, r_{t+1})$ as an example, based on the optimal Bellman equation, use $U_t^{(q)} = r_t + \gamma v(s_{t+1})$ as the true value of s_t, and the $q_i(s_t, a_2)$ value (where i = 0, 1) actually adopted as a predicted value estimate of state s_t, and finally takes MSELoss as the loss function, training the neural network (MSELoss means pair from experience The pool buffer takes a batch data for average work). which is:

$$\boldsymbol{Loss} = frac1\beta(st, at, rt + 1, st + 1)[qist, at; wi - Ut(q)]2 \tag{9}$$

Table 1. Comparison between Soft Actor-Critic (DSAC) with the state-of-the-art models on DomainNet (accuracy %). The source domains and target domain are shown at the top of each column.

Methods	inf, pnt, qdr rel, skt ⟶ clp	clp, pnt, qdr rel, skt ⟶ inf	clp, inf, qdr rel, skt ⟶ pnt	clp, inf, pnt rel, skt ⟶ qdr	clp, inf, pnt qdr, skt ⟶rel	clp, inf, pnt qdr, rel ⟶ skt	Avg
Source only	52.1	23.4	47.7	13.0	60.7	46.5	40.6
ADDA [18]	47.5	11.4	36.7	14.7	49.1	33.5	32.2
MCD [16]	54.3	22.1	45.7	7.69	58.4	43.5	38.5
DANN [4]	60.6	25.8	50.4	7.7	62.0	50.7	43.0
DCTN [22]	48.5	23.4	47.8	7.0	53.4	47.3	38.1
Meta-MCD [6]	62.0	20.4	50.5	15.0	64.3	50.1	43.2
CMSS [24]	64.0	27.8	53.5	15.8	63.3	53.7	46.4
DRT [7]	69.4	30.8	59.17	9.8	66.7	59.2	49.3
DSAC	**71.04**	**32.74**	**59.79**	**15.20**	**69.88**	**60.78**	**51.57**

Q-value Critic Network Update. Take the data from the experience pool buffer $(s_t, a_t, s_{t+1}, r_{t+1})$ to update the Q-value critic network to $(s_t, a_2, s_{t+1}, r_{t+1})$ take an example, with a status value estimate with entropy-containing bonus, that is, the following formula is the true value of the Q-value critic network output:

$$U_t^{(v)} = \sum_{a'_t \epsilon \mathbb{A}(s_t)} \pi(a'_t|s_t;\theta) \left[\min_{i=0,1} q_i \left(s_t, a'_t; w^{(i)} \right) - \alpha \ln \pi \left(a'_t|s_t;\theta \right) \right] \tag{10}$$

The output of the Q-value critic network is used as the predicted value and finally uses MSEloss as the loss function to train the neural network Q-value.

It should be noted that MSEloss is meant to make an average work for data from an experience pool buffer to take a batch, namely:

$$Loss = \frac{1}{|\beta|} \sum_{(s_t,a_t,r_{t+1},s_{t+1}) \in \beta} [v(s_t;w^{(v)}) - U_t^{(v)}]^2 \tag{11}$$

Actor network update loss for Actor network training is slightly complex, its expression is:

$$Loss = -\frac{1}{|\beta|} \sum_{(s_t,a_t,r_{t+1},s_{t+1} \in \beta)} E_{a'_t}[q_0(s_t, a'_t) - \alpha \ln \pi(a'_t \mid s_t;\theta) \tag{12}$$

$E_{a'_t \; \pi(\cdot|s_t;\theta)}[\cdots]$ need to take the item inside the bracket, pay attention to a'_t prime is not in buffer a_t in the extracted data $(s_t, a_t, s_{t+1}, r_{t+1})$, but reusing all possible actions predicted by actor network π, so for discrete action space, often have the following equivalents:

$$
\begin{aligned}
& E_{a'_t \ \pi(\cdot|s_t;\theta)}[q_0(s_t, a'_t; w^{(0)}) - \alpha \ln \pi(a'_t|s_t;\theta)] \\
= & \sum_{a'_t \in \mathbb{A}(s_t)} [q_0(s_t, a'_t; w^{(0)}) - \ln \pi(a_{t+1}|s_t;\theta)
\end{aligned} \tag{13}
$$

The β represents the experience pool buffer, which requires the sample to take out the sample in the experience pool when the loss is required, which can reflect the average meaning of the removal sample. α is the coefficient of entropy, which determines the importance of entropy $\ln \pi(a_{t+1}|s_t;\theta)$, the more important it is.

4 Experiments

This part will evaluate the performance of the enhanced learning selector by experiment.

4.1 Datasets and Experiment Settings

After [13], we consider the data set DomainNet, which contains images from multiple domains. Each domain is alternate as a target domain, and the remaining domains are used as a source domain. All experiments were repeated 5 times and reported mean and variance.

DomainNet: DomainNet is a 600,000 different styles from 645 categories from 6 domains. clipart (clp), infograph (inf), painting (pnt), quickdraw (qdr), real (rel), and sketch (skt). The result was obtained by imagenet [3] pre-trained ResNet-101 [17]. The training of 15 stages of the network, in the initial learning rate is 0.001, the batch size is 64. And the learning rate is 0.1 per five stages.

4.2 Comparison with the Latest Technology

Strengthening the Learning Selector and the results of the DomainNet data set in the literature. In these experiments, the enhanced learning selector is achieved by SAC, achieving loss of loss as a reward.

DomainNet Dataset Evaluation: For DomainNet [13], reset-101 [12] is used as the backbone, and the SAC is compared to 11 baselines. Among them, ADDA [18], DANN [4] and MCD [16] was developed for the traditional non-supervised domain adaptation (UDA), where the single source domain was assumed. The remaining is a multi-source field adaptation method for the domain tag.

Regarding individuality adaptation, the SAC achieves the best performance of all target domains other than "QuickDraw". This can be interpreted by "QuickDraw" and other domains, and the fact that MCD loss is not yetful in this issue. By using M3SDA loss, better results may be in the table.

4.3 Training Efficiency of DSAC

In order to check the training efficiency of DSAC, we are trained by selecting different threshold.

Table 2. Training efficiency of DSAC.

Methods	Select data	Time (min)	AUC (%)
$DSAC_{0.0}$	7000M	181	69.79
$DSAC_{0.3}$	6300M	164	70.05
$DSAC_{0.5}$	4000M	132	71.06

$SAC_{0.0}$ - To train our DSAC with $\alpha = 0.0$, i.e. use the clip data.

$SAC_{0.3}$ - To train our DSAC with $\alpha = 0.3$, i.e. selecting source data with threshold greater than 0.3.

$SAC_{0.5}$ - To train our DSAC method by setting $\alpha = 0.5$.

As shown in Table 2, we check the number of data instances selected from the source domain to the target domain data set to obtain a good AUC. Obviously, $SAC_{0.0}$ requires more data and time to get higher AUC. Because the distribution of source domain data and target domain data is different. The larger the threshold, the higher the quality of the selected data, the faster the training process. Compared with $SAC_{0.0}$, our $SAC_{0.5}$ spends less time. 57%of the data, so it reaches higher AUC, which indicates that our method is effective.

4.4 Source to Single-Target Adaptation

DomainNet [13] also evaluated the performance of the DSAC on traditional domain adaptation issues (single source domain). In this case, adaptive is performed from each of the other five domains (sources) for each target domain. For each target domain, the average and optimal performance is shown in the table are shown. The DSAC is again significantly superior to the previous domain adaptation method. For example, when "clipart" is a target domain, the average adaptation performance is higher than the average adaptability of the MCD method. On average, in all source and target domains, it is superior to MCD to more than 13.07%. These results show that for hundreds of classes, the dynamic residual transmission can result in very good adaptability even in traditional domain adaptation settings. This confirms that even these problems often have many subdomains.

Finally, the results of the table show that the DSAC training in multi-source domains has better-performed 8.2% better than the average value of the optimal single source domain transfer (49.3% and 40.6%). This improvement is more than about 2% higher than MCD (16.74% to 7.69%). This suggests that the domain adaptability is improved in consideration of various source domains, especially when DSAC is used. The main advantage of the enhanced learning selector is that it can be applied to all settings because it does not require a domain tag. If the problem is a single source or multi-source, its universality makes it insignificant.

5 Conclusion

In this paper, we adapt to the introduction of the reinforce learning selector for the multi-source domain, where the selector uses the SAC algorithm to train. It reduces the burden on the sample and unifies the plurality of source domains into a single source domain, which simplifies the alignment between the source domain and the target domain. The experimental results show that the DSAC achieves better adaptability than the latest methods adapted to multiple source domains. We hope this article can provide a new understanding of multi-source domain adaptation.

Acknowledgements. This work was supported by National Key R&D Program of China (No.2017YFE0116800),National Natural Science Foundation of China(Grant No.U20B2074,U1909202) and supported by Key Laboratory of Brain Machine Collaborative Intelligence of Zhejiang Province(2020E10010).

References

1. Blitzer, J., Crammer, K., Kulesza, A., Pereira, F., Wortman, J.: Learning bounds for domain adaptation. In: Platt, J., Koller, D., Singer, Y., Roweis, S. (eds.) Advances in Neural Information Processing Systems, vol. 20. Curran Associates, Inc. (2007). https://proceedings.neurips.cc/paper/2007/file/42e77b63637ab381e8be5f8318cc28a2-Paper.pdf
2. Bollegala, D., Weir, D., Carroll, J.A.: Using multiple sources to construct a sentiment sensitive thesaurus for cross-domain sentiment classification. In: Proceedings of the 49th Annual Meeting of the Association for Computational Linguistics: Human Language Technologies, pp. 132–141 (2011)
3. Deng, J., Dong, W., Socher, R., Li, L.J., Li, K., Fei-Fei, L.: Imagenet: a large-scale hierarchical image database. In: 2009 IEEE Conference on Computer Vision and Pattern Recognition, pp. 248–255. IEEE (2009)
4. Ganin, Y., Lempitsky, V.: Unsupervised domain adaptation by backpropagation. In: International Conference on Machine Learning, pp. 1180–1189. PMLR (2015)
5. LeCun, Y., Bottou, L., Bengio, Y., Haffner, P.: Gradient-based learning applied to document recognition. Proc. IEEE **86**(11), 2278–2324 (1998)
6. Li, D., Hospedales, T.: Online meta-learning for multi-source and semi-supervised domain adaptation. In: Vedaldi, A., Bischof, H., Brox, T., Frahm, J.-M. (eds.) ECCV 2020. LNCS, vol. 12361, pp. 382–403. Springer, Cham (2020). https://doi.org/10.1007/978-3-030-58517-4_23
7. Li, Y., Yuan, L., Chen, Y., Wang, P., Vasconcelos, N.: Dynamic transfer for multi-source domain adaptation. In: Proceedings of the IEEE/CVF Conference on Computer Vision and Pattern Recognition, pp. 10998–11007 (2021)
8. Li, Z., Zhao, Z., Guo, Y., Shen, H., Ye, J.: Mutual learning network for multi-source domain adaptation. arXiv preprint arXiv:2003.12944 (2020)
9. Liu, M., Han, J., Zhang, H., Song, Y.: Domain adaptation for disease phrase matching with adversarial networks. In: Proceedings of the BioNLP 2018 workshop, pp. 137–141 (2018)
10. Maria Carlucci, F., Porzi, L., Caputo, B., Ricci, E., Rota Bulo, S.: Autodial: Automatic domain alignment layers. In: Proceedings of the IEEE International Conference on Computer Vision (ICCV), pp. 5067–5075 (2017)

11. Matasci, G., Volpi, M., Kanevski, M., Bruzzone, L., Tuia, D.: Semisupervised transfer component analysis for domain adaptation in remote sensing image classification. IEEE Trans. Geosci. Remote Sens. **53**(7), 3550–3564 (2015)
12. Pan, J., Hu, X., Li, P., Li, H., He, W., Zhang, Y., Lin, Y.: Domain adaptation via multi-layer transfer learning. Neurocomputing **190**, 10–24 (2016)
13. Peng, X., Bai, Q., Xia, X., Huang, Z., Saenko, K., Wang, B.: Moment matching for multi-source domain adaptation. In: Proceedings of the IEEE/CVF International Conference on Computer Vision, pp. 1406–1415 (2019)
14. Peng, X., Usman, B., Kaushik, N., Hoffman, J., Wang, D., Saenko, K.: Visda: the visual domain adaptation challenge. arXiv preprint arXiv:1710.06924 (2017)
15. Saenko, K., Kulis, B., Fritz, M., Darrell, T.: Adapting visual category models to new domains. In: Daniilidis, K., Maragos, P., Paragios, N. (eds.) ECCV 2010. LNCS, vol. 6314, pp. 213–226. Springer, Heidelberg (2010). https://doi.org/10.1007/978-3-642-15561-1_16
16. Saito, K., Watanabe, K., Ushiku, Y., Harada, T.: Maximum classifier discrepancy for unsupervised domain adaptation. In: Proceedings of the IEEE Conference on Computer Vision and Pattern Recognition, pp. 3723–3732 (2018)
17. Sun, B., Saenko, K.: Deep CORAL: correlation alignment for deep domain adaptation. In: Hua, G., Jégou, H. (eds.) ECCV 2016. LNCS, vol. 9915, pp. 443–450. Springer, Cham (2016). https://doi.org/10.1007/978-3-319-49409-8_35
18. Tzeng, E., Hoffman, J., Saenko, K., Darrell, T.: Adversarial discriminative domain adaptation. In: Proceedings of the IEEE Conference on Computer Vision and Pattern Recognition (CVPR), pp. 7167–7176 (2017)
19. Wang, B., et al.: A minimax game for instance based selective transfer learning. In: Proceedings of the 25th ACM SIGKDD International Conference on Knowledge Discovery & Data Mining, pp. 34–43 (2019)
20. Wang, H., Xu, M., Ni, B., Zhang, W.: Learning to combine: knowledge aggregation for multi-source domain adaptation. In: Vedaldi, A., Bischof, H., Brox, T., Frahm, J.-M. (eds.) ECCV 2020. LNCS, vol. 12353, pp. 727–744. Springer, Cham (2020). https://doi.org/10.1007/978-3-030-58598-3_43
21. Xia, R., Zong, C., Hu, X., Cambria, E.: Feature ensemble plus sample selection: domain adaptation for sentiment classification. IEEE Intell. Syst. **28**(3), 10–18 (2013)
22. Xu, R., Chen, Z., Zuo, W., Yan, J., Lin, L.: Deep cocktail network: multi-source unsupervised domain adaptation with category shift. In: Proceedings of the IEEE Conference on Computer Vision and Pattern Recognition, pp. 3964–3973 (2018)
23. Yang, J., Yan, R., Hauptmann, A.G.: Cross-domain video concept detection using adaptive SVMS. In: Proceedings of the 15th ACM International Conference on Multimedia, pp. 188–197 (2007)
24. Yang, L., Balaji, Y., Lim, S.-N., Shrivastava, A.: Curriculum manager for source selection in multi-source domain adaptation. In: Vedaldi, A., Bischof, H., Brox, T., Frahm, J.-M. (eds.) ECCV 2020. LNCS, vol. 12359, pp. 608–624. Springer, Cham (2020). https://doi.org/10.1007/978-3-030-58568-6_36

Transfer Learning to Decode Brain States Reflecting the Relationship Between Cognitive Tasks

Youzhi Qu[1], Xinyao Jian[2], Wenxin Che[1], Penghui Du[1], Kai Fu[1], and Quanying Liu[1](✉)

[1] Shenzhen Key Laboratory of Smart Healthcare Engineering, Department of Biomedical Engineering, Southern University of Science and Technology, Shenzhen 518055, China
liuqy@sustech.edu.cn

[2] Department of Statistics and Data Science, Southern University of Science and Technology, Shenzhen 518055, China

Abstract. Transfer learning improves the performance of the target task by leveraging the data of a specific source task: the closer the relationship between the source and the target tasks, the greater the performance improvement by transfer learning. In neuroscience, the relationship between cognitive tasks is usually represented by similarity of activated brain regions or neural representation. However, no study has linked transfer learning and neuroscience to reveal the relationship between cognitive tasks. In this study, we propose a transfer learning framework to reflect the relationship between cognitive tasks, and compare the task relations reflected by transfer learning and by the overlaps of brain regions (*e.g.*, neurosynth). Our results of transfer learning create *cognitive taskonomy* to reflect the relationship between cognitive tasks which is well in line with the task relations derived from neurosynth. Transfer learning performs better in task decoding with fMRI data if the source and target cognitive tasks activate similar brain regions. Our study uncovers the relationship of multiple cognitive tasks and provides guidance for source task selection in transfer learning for neural decoding based on small-sample data.

Keywords: Transfer learning · Task relationship · Cognitive tasks

1 Introduction

Transfer learning leverages the knowledge in the source domain data to transfer to the target domain with an assumption that the source and target tasks in the model share some common knowledge [30,36]. A model pre-trained with source domain data acquires rich high-order knowledge, which can help to learn the target domain task with only a small amount of target domain data. However, transfer learning is not always beneficial. The performance of transfer learning greatly

X. Ying (Ed.): HBAI 2022, CCIS 1692, pp. 110–122, 2023.
https://doi.org/10.1007/978-981-19-8222-4_10

depends on the relationship between tasks in source and target domains. If great distinction exists between the two tasks, transfer learning may negatively affect the learning of target domain, which is also called *negative transfer*. Therefore, it is essential to examine the relationship between tasks in transfer learning. [35] proposes a computational model to calculate the task affinity matrix by comparing the performance of transfer learning between tasks in computer vision (*e.g.*, such as object recognition, depth estimation and edge detection), which provides guidance on how to select source tasks. This is called *taskonomy* [35]. Although the taskonomy in computer vision has been intensively studied and utilized in transfer learning, the relationship between cognitive tasks is less clear. There are no studies using transfer learning to explore the relationship between cognitive tasks.

Exploring the neural mechanisms of information process in cognitive tasks is critical to the understanding of the brain. Some tasks may activate overlapping brain regions or induce similar brain activity. Thus, the task relations can be reflected at the neural level. For example, the fusiform gyrus is involved in multiple tasks, including recognizing faces and understanding the meaning of written words. The activated brain regions when performing tasks with a closer relation have more overlaps than those with less relation, since the former shares neural circuits in cognitive process [21]. This strategy of our brain evolves to improve multitasking and efficiency [19]. From a neuroscience perspective, the relationship between cognitive tasks can be reflected in brain region overlap [20], or in neural representation similarity [18,25]. In the era of deep learning, attention has been paid to the similarities and differences between the brain and artificial intelligence (AI) in processing information [9,13]. Neural representation similarity in artificial neurons has been also applied to reflect the relations of task stimulus, especially visual stimuli [28,32]. Although some studies have found similarities in neural representations of the brain and AI [22,33]; however, few studies have focused on the commonalities in task performance between the human brain and AI.

In this study, we present a transfer learning framework to create the relationship between cognitive tasks, called *cognitive taskonomy* (Fig. 1). It obtains a task affinity matrix to represent the relations of various cognitive tasks. The task affinity matrices from transfer learning and from brain activity are compared to examine the resemblance of cognitive taskonomy from AI and from the human brain. Our contributions are summarized as follows.

- We propose a computational modeling framework of cognitive task relations and create the cognitive taskonomy using transfer learning (Fig. 1).
- By comparing the cognitive task affinity matrix derived from transfer learning and from Neurosynth, we uncover a strong resemblance of these two, especially in the emotion, gambling and social tasks (Fig. 4).
- The affinity of seven cognitive tasks provides guidance for source task selection in transfer learning for brain state decoding (Fig. 5 & 6).

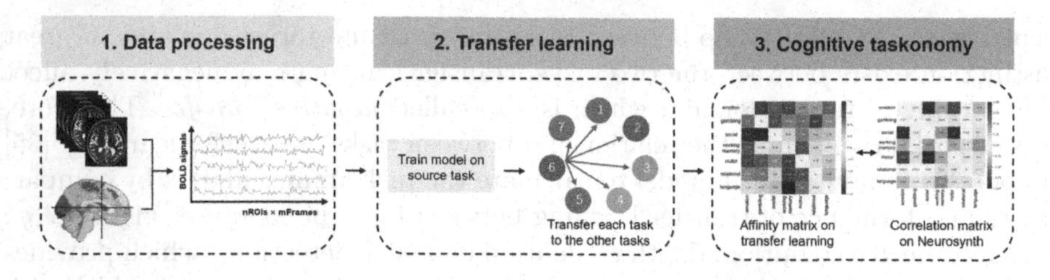

Fig. 1. Computational modeling of cognitive task relations and creating the *cognitive taskonomy* using transfer learning. From left to right: (1) data preprocessing: preprocessing fMRI data and parcellating brain with atlas; (2) transfer learning: training task-specific networks for task decoding and transferring the network trained with the source task to the rest six tasks; (3) cognitive taskonomy: comparing task affinity matrices derived from transfer learning and Neurosynth.

2 Related Work

2.1 Cognitive Task Relations from Neuroscience Perspective

Different cognitive tasks involve a variety of cognitive processes and brain functions. From the neuroscience perspective, the relationship between cognitive tasks can be investigated at different levels, including the overlaps of activated brain regions, the neural representation similarity and the representational similarity in neural decoding models. Similar cognitive processes activate some shared brain regions, so the overlap of activated brain regions can be used to measure the relationship between tasks. Previous studies have shown that the activation of brain regions exhibits a community structure under cognitive task states, and the overlapping part of the community reflects the functional relationship of cognitive tasks [20,31]. Moreover, the similarity of neural representations or the representational similarity of neural decoding models can also reflect the relationship between different cognitive tasks. A large number of neuroscience studies aim to explore the relationship between stimuli and neural responses. For instance, in a vision task, a mapping between facial features and neural representations is constructed [18]. Recently, artificial neural networks have been employed as encoding models to construct the mapping from the stimuli to neural responses [2,22] and as decoding models to classify cognitive states from neural data. Representations of artificial neurons in decoding models are used to measure the similarity of cognitive tasks [14,16,33].

2.2 Cognitive Task Relations From Transfer Learning Perspective

Transfer learning seeks to utilize the knowledge of a source task to a target task. We define a domain in transfer learning as $\mathcal{D} = \{\mathcal{X}, P(X)\}$, where $\mathcal{X}$ is the feature space and $P(X)$ is a marginal probability distribution, $X = \{x_1, ..., x_n\}$ is sampled from $\mathcal{X}$. In fMRI experiments, $\mathcal{X}$ includes all possible images collected under the same experiment protocol, and $P(X)$ depends on subject groups, such as children or adults. A task is defined as a combination of a label space $\mathcal{Y}$ and a predictive function $f(\cdot)$, i.e. $\mathcal{T} = \{\mathcal{Y}, f(\cdot)\}$. The predictive function f is learned from the training data $\{(x_i, y_i)\}_{i=1}^{n}$, where $x_i \in \mathcal{X}$ and $y_i \in \mathcal{Y}$. Given a source domain $\mathcal{D}_S = \{\mathcal{X}_S, P(X_S)\}$ and task $\mathcal{T}_S = \{\mathcal{Y}_S, f_S(\cdot)\}$, and a target domain $\mathcal{D}_T = \{\mathcal{X}_T, P(X_T)\}$ and task $\mathcal{T}_T = \{\mathcal{Y}_T, f_T(\cdot)\}$, our purpose is to improve the generalization of the target predictive function f_T in $\mathcal{T}_T$ by utilize the knowledge we learned from $\mathcal{D}_S$ and $\mathcal{T}_S$.

Ideally, we expect that our knowledge acquired from the source domain improves the performance of the target predictive function $f_T(\cdot)$, which is called *transferability.* The transferability is largely decided by the relationship of data in the source and target domains, as well as the source and target tasks. However, in transfer learning, models trained on the some tasks may not be able to be transferred to new tasks. The difference between the source and target domains can have a negative impact when performing transfer learning. For example, [23] experimentally shows that if two tasks are too dissimilar, transfer learning may hurt, rather than improve, the performance of the target task.

There have been some studies using transfer learning to evaluate the relationship between transferability and similarity of source and target tasks. For example, [35] investigated the underlying relational structure of different tasks by computing the affinity matrix based on the ability to solve one task using representations trained for another task. Their results suggest that the higher similarity of representations among tasks leads to better transferability in transfer learning. [1] developed a TASK2Vec method that can provide a fixed-dimensional embedding of the task. They demonstrated that this embedding can predict task similarities that match our intuition about semantic and taxonomic relations between different tasks, and is also helpful for us to choose a pre-trained model to solve a new task. [11] used representation similarity analysis to obtain a similarity score among tasks by computing correlations between models trained on different tasks. Their results reveal that the higher the similarity score between tasks, the better the transfer learning performance between them.

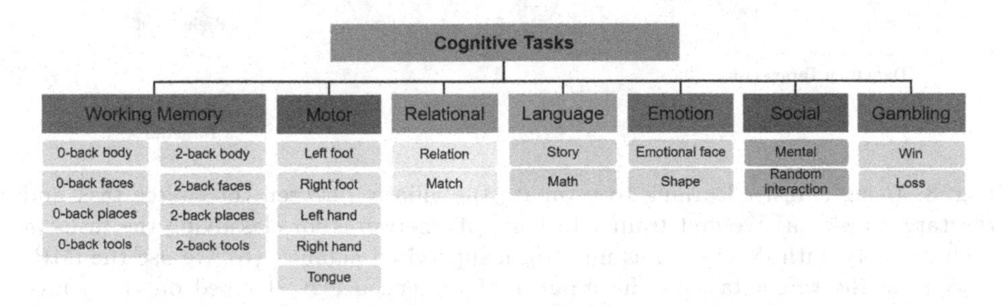

Fig. 2. Task Description. HCP fMRI dataset includes 23 tasks, belonging to 7 categories.

3 Methods

3.1 HCP Tasks

Our experiments are conducted on a large fMRI dataset, *i.e.*, Human Connectome Project (HCP) S1200 Release [27]. The HCP fMRI data are recorded from over 1,000 subjects while they are performing 7 categories of cognitive tasks, including the emotion processing tasks [12], the gambling tasks [8], the social cognition tasks [6], the working memory tasks [10], the motor tasks [5], the relational processing tasks [24] and language processing tasks [3]. Each category consists of 2–8 subtasks (Fig. 2). To eliminate the effect of different number of subtasks on transfer learning, we choose 0-back faces and 2-back faces as subtasks of working memory category, and left hand and right hand as subtasks of motor category. The HCP fMRI data are preprocessed using the standard HCP pipeline for removing spatial distortions, motion correction, registration and normalization. We then parcellate the brain into 90 regions using automatic anatomical labeling (AAL) atlas [26]. The preprocessed HCP task-based fMRI data is detailed in Table 1.

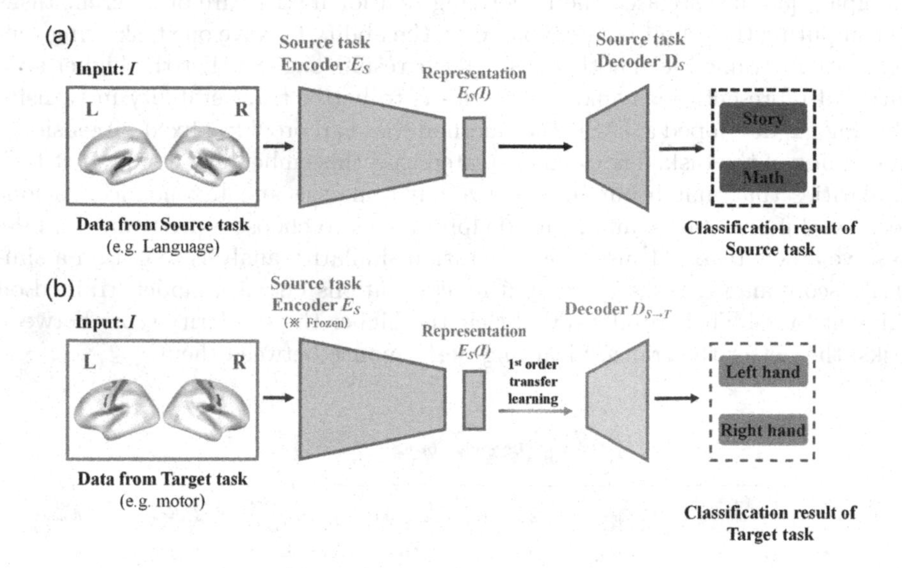

Fig. 3. Using transfer learning to estimate the affinity between the source task and the target task. (a) We first train 7 task-specific networks for classifying the tasks in each category with fMRI data as input in a supervised manner. (b) We use the fMRI data from the target task as the input I of the encoder E_S trained on the source task S. The parameters of the encoder are frozen during transfer learning. We train a decoder $D_{S\to T}$ with the representation $E_S(I)$ in the encoder as the input to perform classification on the target task T. The classification accuracy from 7 source tasks to 7 target tasks is the 7×7 task affinity matrix.

Table 1. Details of HCP task-based fMRI data in our study.

Category of tasks	# of subjects	# of subtasks for classification
Working memory	1077	2
Motor	1076	2
Relational processing	1036	2
Language processing	1040	2
Emotion processing	1040	2
Social cognition	1044	2
Gambling	1080	2

3.2 Transfer Learning

For each category of tasks, we first train a 4-layer fully connected neural network with fMRI data as the input to perform classification among subtasks. Then, we transfer the seven task-specific pre-trained neural networks to the target domain task, resulting in a *task affinity matrix* to reflect the relations of cognitive tasks (Fig. 3).

Network Architecture: There are a total of seven task-specific networks for classification tasks, which correspond to 7 cognitive categories. For each classification task, the classification model should identify the specific task states under the category. The corresponding relationship between categories and tasks is shown in Fig. 2. The classification model is divided into two parts: encoder and decoder. For each task, we use the same model architecture consistent to avoid additional bias. Specifically, both the encoder and decoder consist of 2 fully connected layers.

Training Task-Specific Networks: We first train 7 task-specific networks for classifying the tasks in each category with fMRI data as input in a supervised manner. These 7 task-specific networks are treated as the gold standard of source domain. Specifically, the task-related fMRI data in the source domain is divided into training set, validation set and test set with the ratio of 8:1:1. All models are trained using the Adam optimizer with an initial learning rate of 0.001 and cross-entropy loss. The maximum epoch number is set to 50. When model performance stopped improving on the validation set at 5 epochs, we apply the early stopping criterion to stop training. The performance of decoding cognitive tasks is quantified with classification accuracy.

Transfer Learning: After the gold-standard task-specific neural networks are trained, we perform transfer learning. We treat each task-specific network for classifying one of the 7 categories as the source task, and the rest 6 categories as the target task separately. In transfer learning, we fix the parameters of the two fully connected layers of the encoder in the model, and train the model with 1%, 5%, 10%, 20%, 30%, 40% and 50% proportions of target domain data as training set, respectively. Another 10% proportion of target domain data is used as the validation set, and early stopping is applied to reduce overfitting. For each proportion of the transfer results, we repeat 10 times and average the accuracy.

3.3 Validation of Cognitive Taskonomy

To validate the cognitive taskonomy obtained by transfer learning, we used the task-related brain regions from Neurosynth [34], an fMRI meta-analysis platform. Specifically, we input 7 keywords (*i.e.*, working memory, motor, relational, language, emotions, social cognition, gambling) to Neurosynth for retrieving the brain topological maps corresponding to the 7 categories of tasks. Then, we compute the Pearson correlation of the brain maps, representing the similarity matrix across tasks. We set diagonal of the similarity matrix to 0, and normalize positive and negative values in the matrix to the intervals 0 to 1 and from 0 to –1, respectively.

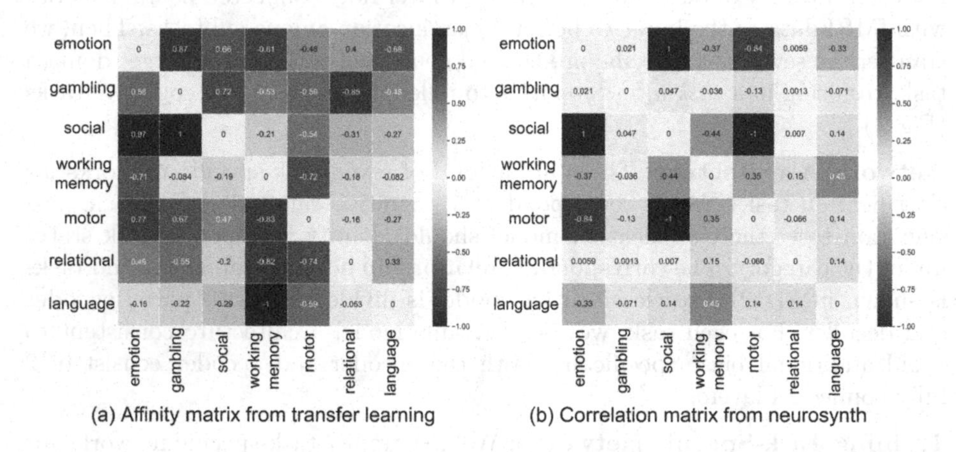

Fig. 4. Cognitive taskonomy. (a) The accuracy change from the source task to the target task using transfer learning. The positive value represents that the source task contributes to improving the accuracy of target task, indicating the similarity between the source and task tasks; vice versa. (b) The correlation of cognitive tasks derived from the brain regions. The task-related brain regions are exported from Neurosynth, an online fMRI meta-analysis platform.

4 Results

4.1 Affinity Matrix of Cognitive Tasks

We use the performance of training task-specific supervised networks with 1% target domain data as the baseline, and check whether using different source domain data for transfer learning can improve the performance of brain decoding. Fig. 4 (a) shows the accuracy changed from the source task to the target task between seven categories. The row in figure represents the source task, the column in figure represents the target task. We normalize positive and negative values in the matrix to the intervals 0 to 1 and from 0 to -1, respectively, which is convenient for comparing the relationship between different source tasks and the target tasks. The (i, j)-th element of the affinity matrix in Fig. 4 (a) represents the transferring from the i-th task to the j-th task. For instance, the first three categories of tasks (*i.e.*, emotion, gambling, and social tasks) show positive transfer to each other. In contrast, they have negative transfer to working memory and motor tasks. Fig. 4 (b) show the Pearson correlation coefficient of the brain topological maps from Neurosynth. The results show that this affinity matrix is high resemble of the task affinity matrix from transfer learning, especially the emotion, gambling and social tasks. Moreover, the tasks with more overlaps in brain regions (*e.g.*, emotion, gambling and social) can better improve the brain decoding performance of the target tasks in transfer learning, while the negative correlation in activity maps brings negative transfer (*e.g.*, emotion to working memory, emotion to motor).

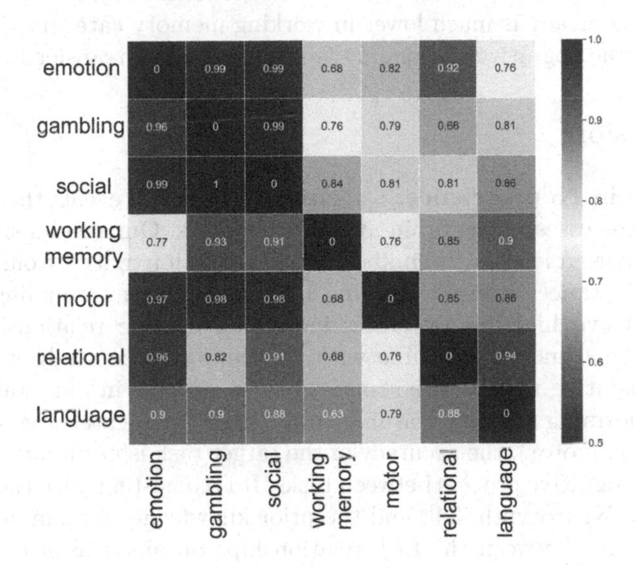

Fig. 5. The accuracy comparison of transfer learning with 1% data and supervised learning with 80% data. The value represents the ratio of transfer learning performance to supervised learning performance. The larger the value, the closer the performance of transfer learning is to the gold standard.

4.2 Compare with Task-Specific Networks

We use the performance of training task-specific supervised networks with 80% target domain data as the gold standard. We compare the brain decoding performance of the gold standard with transfer learning using 1% target domain data. Fig. 5 shows the ratio of the decoding performance of transfer learning to the gold standard. The larger the ratio, the closer the transfer performance is to the task-specific network performance. The result shows that when the three categories, emotion, gambling and social are used as the target tasks, the transfer learning from source task for other categories can achieve the performance close to the gold standard when using a small amount of data. However, when the working memory is used as target tasks, the effect of transfer learning is not good.

4.3 Brain Decoding Accuracy with Transfer Learning

Fig. 6 shows the performance of transfer learning between 7 categories when using various proportions of target domain data ranging from 1% to 50%. The brain decoding accuracy of the target task increases with the amount of target domain data. No matter what the source task is, transfer learning on three target domain tasks (*i.e.*, emotion, gambling and social) is consistently higher than other target tasks. Remarkably, the classification accuracy achieves over 90% in these four tasks with only 1% of target domain data. In contrast, the classification accuracy is much lower in working memory category. These results indicate that the cognitive taskonomy is informative for transfer learning.

5 Discussion

Significance in Neuroscience: Cognitive taskonomy reveals the relationship of cognitive processes in the brain at cognitive tasks. Our results show the task affinity matrix from learning is similar to the correlation matrix from Neurosynth (Fig. 4). For instance, transfer learning between emotion, gambling and social categories achieve high performance, indicating a close relationship between these tasks. This finding is in line with the neuroscience study which reports that these cognitive tasks share some cognitive process in our brain [4,15]. In contrast, performing transfer learning from these three categories to the other categories (*e.g.*, motor), the accuracy of the target task is not improved, suggesting a distinct cognitive process between tasks. It is consistent with the brain maps exported from Neurosynth [34], and the prior knowledge of brain networks [17]. This resemblance between the task relationships obtained from transfer learning's performance and the task relationships obtained from brain regions implies that AI and the brain are coherent in processing cognitive tasks.

Significance in Transfer Learning: Cognitive taskonomy is informative for the selection of source tasks in transfer learning for neural decoding applications, especially when the training data is extremely small. Transfer learning performance between 7 cognitive categories (Fig. 5) indicates that when the emotion, gambling and social cognitive categories are used as source tasks to perform transfer learning on other categories, the accuracy of the target tasks is close to the gold standard, even using only 1% target domain data. However, when the working memory cognitive category is used as source task, the accuracy of the target tasks is poor. The results indicate that the emotion, gambling and social cognitive categories are transferable, while the working memory [29] cognitive category is non-transferable. Fig. 6 shows that the transferability and

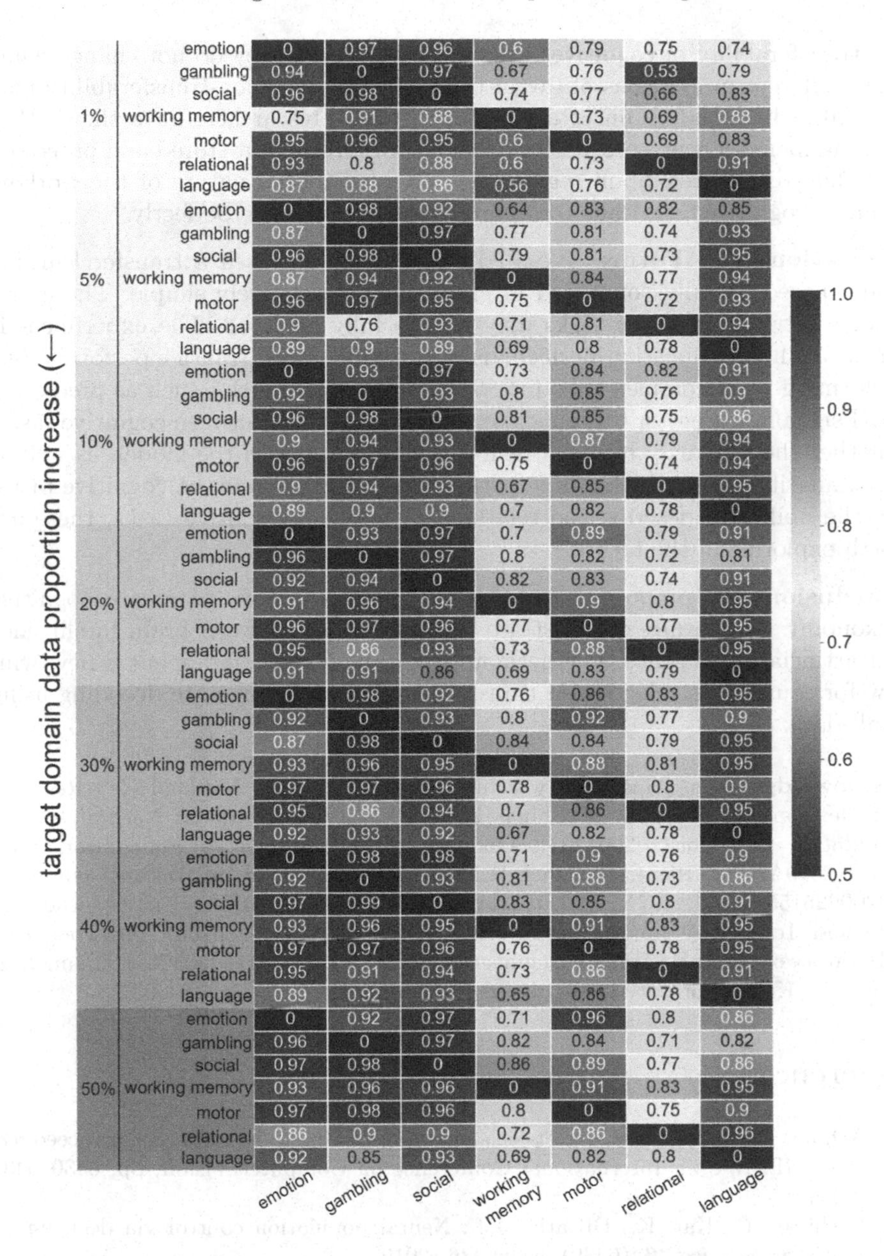

Fig. 6. The performance in transfer learning. The model is trained using 1%, 5%, 10%, 20%, 30%, 40% and 50% of target domain data, respectively. The category in the row represents the source task and the category in the column represents the target task. The values represent the accuracy of the transfer learning model on target tasks in test sets.

non-transferability of cognitive tasks are robust, and they do not change when using different proportions of target domain data. This non-transferability may be related to decoding uncertainty. For instance, researchers have shown that working memory is not processed in a single brain site, but stored and processed in widely distributed brain regions [7]. The distributed nature of the working memory cognitive task leads to a hard problem to decode perfectly.

Limitations and Future Work: In this study, we used a transfer learning model based on fully connected layers which are relatively simple. The correlations between cognitive tasks may not be fully explored. The experiment is based on the classification performances to evaluate the transfer performances. Performing more complex decoding tasks on cognitive tasks, such as predicting fMRI signals, is a better way to explore the relationship between cognitive tasks. Whether the similarity between neural representations in the model, as well as the similarity in transfer learning performance changes between cognitive tasks, has the ability to describe the relationship between cognitive tasks, these are worth exploring in the future.

Conclusion: We propose a transfer learning framework to create cognitive taskonomy. The results demonstrate the similarities between brain intelligence and artificial intelligence. Furthermore, cognitive taskonomy opens a new window for source task selection in transfer learning for brain state decoding using small data.

Acknowledgments. This work was funded in part by the National Key Research and Development Program of China (2021YFF1200804), National Natural Science Foundation of China (62001205), Guangdong Natural Science Foundation Joint Fund (2019A1515111038), Shenzhen Science and Technology Innovation Committee (20200925155957004, KCXFZ2020122117340001), Shenzhen-Hong Kong-Macao Science and Technology Innovation Project (SGDX2020110309280100), Shenzhen Key Laboratory of Smart Healthcare Engineering (ZDSYS20200811144003009), Guangdong Provincial Key Laboratory of Advanced Biomaterials (2022B1212010003).

References

1. Achille, A., et al.: Task2vec: task embedding for meta-learning. In: Proceedings of the IEEE/CVF International Conference on Computer Vision, pp. 6430–6439 (2019)
2. Bashivan, P., Kar, K., DiCarlo, J.J.: Neural population control via deep image synthesis. Science **364**(6439), eaav9436 (2019)
3. Binder, J.R., Gross, W.L., Allendorfer, J.B., Bonilha, L., Chapin, J., Edwards, J.C., Grabowski, T.J., Langfitt, J.T., Loring, D.W., Lowe, M.J., et al.: Mapping anterior temporal lobe language areas with FMRI: a multicenter normative study. Neuroimage **54**(2), 1465–1475 (2011)
4. Britton, J.C., Phan, K.L., Taylor, S.F., Welsh, R.C., Berridge, K.C., Liberzon, I.: Neural correlates of social and nonsocial emotions: an FMRI study. Neuroimage **31**(1), 397–409 (2006)

5. Buckner, R.L., Krienen, F.M., Castellanos, A., Diaz, J.C., Yeo, B.T.: The organization of the human cerebellum estimated by intrinsic functional connectivity. J. Neurophysiol. **106**(5), 2322–2345 (2011)
6. Castelli, F., Happé, F., Frith, U., Frith, C.: Movement and mind: a functional imaging study of perception and interpretation of complex intentional movement patterns. Neuroimage **12**(3), 314–325 (2000)
7. Christophel, T.B., Klink, P.C., Spitzer, B., Roelfsema, P.R., Haynes, J.D.: The distributed nature of working memory. Trends Cognit. Sci. **21**(2), 111–124 (2017)
8. Delgado, M.R., Nystrom, L.E., Fissell, C., Noll, D., Fiez, J.A.: Tracking the hemodynamic responses to reward and punishment in the striatum. J. Neurophysiol. **84**(6), 3072–3077 (2000)
9. DiCarlo, J., Zoccolan, D., Rust, N.: How does the brain solve visual object recognition? Neuron **73**, 415–434 (2012)
10. Drobyshevsky, A., Baumann, S.B., Schneider, W.: A rapid FMRI task battery for mapping of visual, motor, cognitive, and emotional function. Neuroimage **31**(2), 732–744 (2006)
11. Dwivedi, K., Roig, G.: Representation similarity analysis for efficient task taxonomy & transfer learning. In: Proceedings of the IEEE/CVF Conference on Computer Vision and Pattern Recognition, pp. 12387–12396 (2019)
12. Hariri, A.R., Brown, S.M., Williamson, D.E., Flory, J.D., De Wit, H., Manuck, S.B.: Preference for immediate over delayed rewards is associated with magnitude of ventral striatal activity. J. Neurosci. **26**(51), 13213–13217 (2006)
13. Klindt, D.A., Ecker, A.S., Euler, T., Bethge, M.: Neural system identification for large populations separating what and where. In: Advances in Neural Information Processing Systems (NeurIPS), pp. 3509–3519 (2017)
14. Li, H., Fan, Y.: Interpretable, highly accurate brain decoding of subtly distinct brain states from functional MRI using intrinsic functional networks and long short-term memory recurrent neural networks. NeuroImage **202**, 116059 (2019)
15. Li, X., Lu, Z.L., D'Argembeau, A., Ng, M., Bechara, A.: The IOWA gambling task in FMRI images. Hum. Brain Mapping **31**(3), 410–423 (2010)
16. Li, X., et al.: Braingnn: interpretable brain graph neural network for FMRI analysis. Med. Image Anal. **74**, 102233 (2021)
17. Liu, Q., Farahibozorg, S., Porcaro, C., Wenderoth, N., Mantini, D.: Detecting large-scale networks in the human brain using high-density electroencephalography. Hum. Brain Mapping **38**(9), 4631–4643 (2017)
18. Loffler, G., Yourganov, G., Wilkinson, F., Wilson, H.R.: FMRI evidence for the neural representation of faces. Nat. Neurosci. **8**(10), 1386–1391 (2005)
19. Marois, R., Ivanoff, J.: Capacity limits of information processing in the brain. Trends Cogn. Sci. **9**(6), 296–305 (2005)
20. Najafi, M., McMenamin, B.W., Simon, J.Z., Pessoa, L.: Overlapping communities reveal rich structure in large-scale brain networks during rest and task conditions. Neuroimage **135**, 92–106 (2016)
21. Purves, D., Augustine, G.J., Fitzpatrick, D., Hall, W., LaMantia, A.S., White, L.: Neurosciences. De Boeck Supérieur (2019)
22. Ran, X., et al.: Deep auto-encoder with neural response. arXiv preprint arXiv:2111.15309 (2021)
23. Rosenstein, M.T., Marx, Z., Kaelbling, L.P., Dietterich, T.G.: To transfer or not to transfer. In: In NIPS'05 Workshop, Inductive Transfer: 10 Years Later (2005)
24. Smith, R., Keramatian, K., Christoff, K.: Localizing the rostrolateral prefrontal cortex at the individual level. Neuroimage **36**(4), 1387–1396 (2007)

25. Striem-Amit, E., Wang, X., Bi, Y., Caramazza, A.: Neural representation of visual concepts in people born blind. Nat. Commun. **9**(1), 1–12 (2018)
26. Tzourio-Mazoyer, N., Landeau, B., Papathanassiou, D., Crivello, F., Etard, O., Delcroix, N., Mazoyer, B., Joliot, M.: Automated anatomical labeling of activations in SPM using a macroscopic anatomical parcellation of the MNI MRI single-subject brain. Neuroimage **15**(1), 273–289 (2002)
27. Van Essen, D.C., et al.: The wu-minn human connectome project: an overview. Neuroimage **80**, 62–79 (2013)
28. Walker, E.Y., et al.: Inception loops discover what excites neurons most using deep predictive models. Nat. Neurosci. **22**(12), 2060–2065 (2019)
29. Wang, X., et al.: Decoding and mapping task states of the human brain via deep learning. Hum. Brain Mapp. **41**(6), 1505–1519 (2020)
30. Weiss, K., Khoshgoftaar, T.M., Wang, D.: A survey of transfer learning. J. Big data **3**(1), 1–40 (2016)
31. Wu, H., Mai, X., Tang, H., Ge, Y., Luo, Y.J., Liu, C.: Dissociable somatotopic representations of Chinese action verbs in the motor and premotor cortex. Sci. Rep. **3**(1), 1–12 (2013)
32. Yamins, D., DiCarlo, J.: Using goal-driven deep learning models to understand sensory cortex. Nat. Neurosci. **19**, 356–365 (2016)
33. Yang, G.R., Joglekar, M.R., Song, H., Newsome, W., Wang, X.: Task representations in neural networks trained to perform many cognitive tasks. Nat. Neurosci. **22**, 297–306 (2019)
34. Yarkoni, T., Poldrack, R.A., Nichols, T.E., Van Essen, D.C., Wager, T.D.: Large-scale automated synthesis of human functional neuroimaging data. Nat. Meth. **8**(8), 665–670 (2011)
35. Zamir, A.R., Sax, A., Shen, W., Guibas, L.J., Malik, J., Savarese, S.: Taskonomy: Disentangling task transfer learning. In: Proceedings of the IEEE Conference on Computer Vision and Pattern Recognition, pp. 3712–3722 (2018)
36. Zhuang, F., et al.: A comprehensive survey on transfer learning. Proc. IEEE **109**(1), 43–76 (2020)

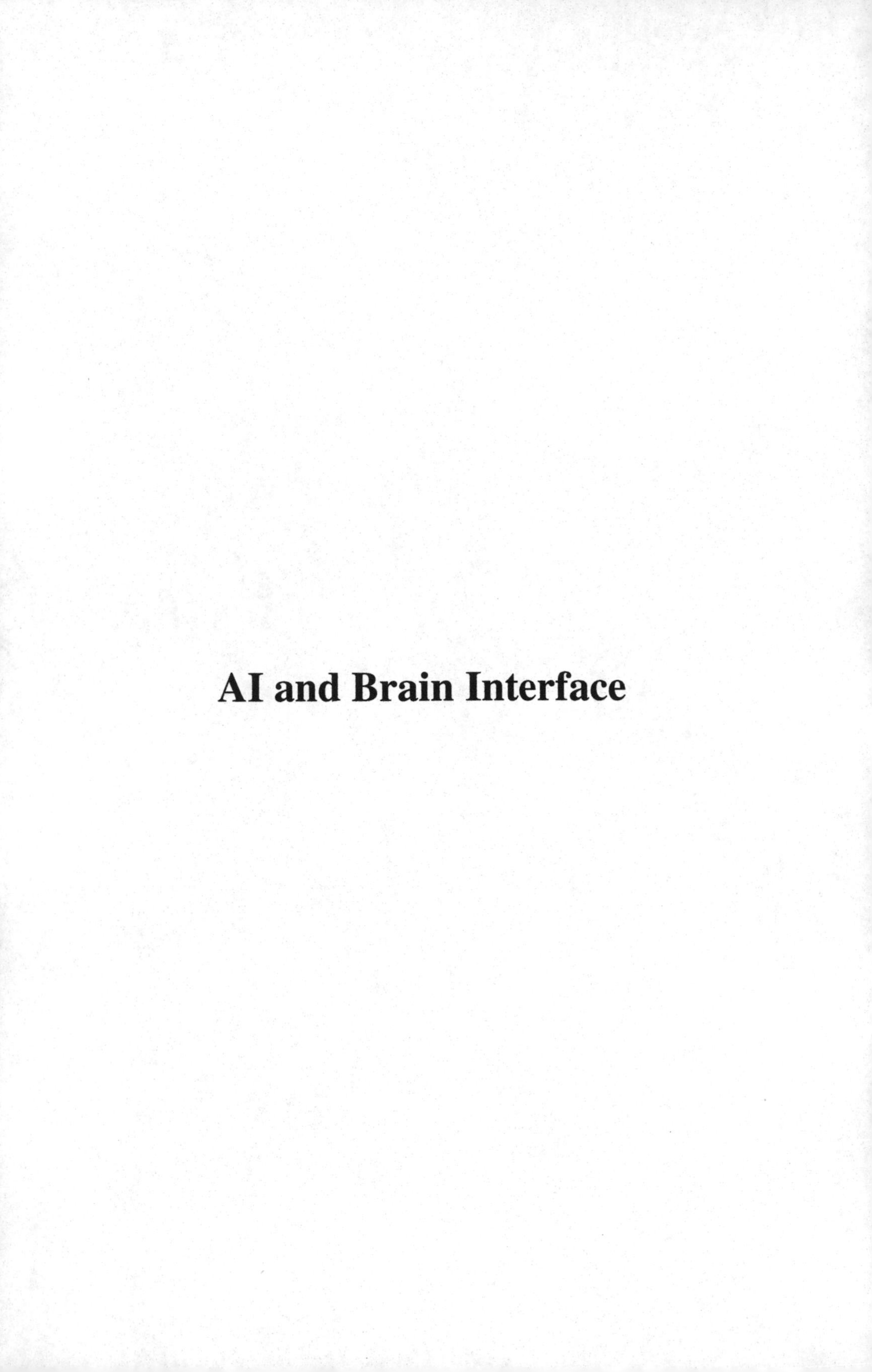

AI and Brain Interface

Brain Network Analysis of Hand Motor Execution and Imagery Based on Conditional Granger Causality

Yuqing He[1], Bin Hao[2(✉)], Abdelkader Nasreddine Belkacem[3], Jiaxin Zhang[1], Penghai Li[1], Jun Liang[4], Changming Wang[5], and Chao Chen[1]

[1] Tianjin University of Technology, Tianjin 300384, China
lph1973@tju.edu.cn
[2] Zhonghuan Information College Tianjin University of Technology, Tianjin 300380, China
haobin1985@163.com
[3] Department of Computer and Network Engineering, College of Information Technology, UAE University, P.O. Box 15551, Al Ain, UAE
belkacem@uaeu.ac.ae
[4] Tianjin Medical University General Hospital, Tianjin 300052, China
Evanliangjun@tmu.edu.cn
[5] Beijing Key Laboratory of Mental Disorders, Beijing Anding Hospital, Capital Medical University, Beijing 100088, China

Abstract. The exploration of neural activity patterns in motor imagery offers a new way of thinking for improving motor skills in normal individuals and for rehabilitating patients with motor disorders. In this paper, the influence relationship between the brain network of the brain motor system and the relevant motor intervals was investigated by collecting EEG signals during finger motor execution and motor imagery from 11 subjects. To address the problem that Granger causality can only reflect the interaction between two temporal variables, a conditional Granger causality analysis was introduced to analysis the brain network relationships between multiple motor compartments. The results showed that the brain network map of finger motor execution had more effective connections than that of finger motor imagination, and it was found that there were effective connection loops between left PMA and left MA, left MA and left SA, and left SA and right SA for both finger motor execution and motor imagination, and the most important connection in motor function was from premotor area to primary motor area.

Keywords: Motor execution · Motor imagery · EEG signal · Conditional granger causality analysis · Brain network

1 Introduction

Motor imagery plays an important role in the neural remodeling of damaged nerves, and during motor execution and motor imagery, there are a large number of overlapping brain

X. Ying (Ed.): HBAI 2022, CCIS 1692, pp. 125–134, 2023.
https://doi.org/10.1007/978-981-19-8222-4_11

areas that are activated by both, such as primary motor area (MA), premotor area (PMA), primary sensory area (SA), etc. [1]. Numerous studies have concluded that the premotor and primary sensory areas play a key role in motor execution and motor preparation, and therefore it is inferred that they also play an important role in motor imagery [2–4].

In a comparative analysis of motor execution and motor imagery, researchers found that although the brain areas activated by the two processes are similar and there are many overlapping brain areas, the brain network connections are more complex during motor execution [5, 6]. Therefore, this paper proposes to study the brain network of motor imagery and compare it with the brain network of motor execution, and it is of great importance to study and analyze the differences.

Most of the current literature on the use of brain net-works or effective connectivity to study motor imagery has been analyzed using fMRI techniques, and research on the analysis of EEG techniques is just beginning. Friston et al. [7, 8] successively proposed functional connectivity and effective connectivity, which became a theoretical basis for researchers to study brain networks. Solodkin [9] selected some representative brain regions of interest, and used structural equation modeling to study and compare the effective connection between motor execution and motor imagery in hand movement. Two conditions were found in which auxiliary motor areas, the dorsal premotor area and the intraparietal sulcus, had opposite effects on inputs to primary motor areas. In motor imagery since there is no actual movement, certain brain areas act opposite to primary motor areas relative to motor execution, and Solodkin referred to this negative connectivity in motor imagery as inhibition [10]. Wang et al. [11] also found that compared with normal subjects, the two brain regions that are functionally connected are physically farther apart in patients with schizophrenia. More and more studies have shown that analyzing brain network structure can pro-vide a powerful tool for humans to explore the way the brain works, study the pathological mechanism of neurodegenerative diseases, and improve the diagnosis and treatment of mental diseases and brain injuries.

In this paper, the Electroencephalogram (EEG) signals of motor execution and motor imagery of right-handed subjects are first collected, and then the brain network of the whole brain is analyzed. Here, because the brain network nodes of the whole brain contain more than three time series, so the brain network was subsequently analyzed using the improved Conditional Granger causality (CGranger causality) method [12].

2 Methods

2.1 Conditional Granger Causality Analysis

Granger causality is a way to describe the interaction of two signals. Granger causality analysis model is based on an autoregressive model, which relies on the temporal priority of signals and can be used to measure the degree of interaction between signals. It can be used to explore the temporal relationship between regions of interest It can be used to explore the temporal relationships between regions of interest, thus revealing the directional information flow between brain regions. The Granger causal analysis model has been used to calculate brain connectivity for decision making and motor recovery. Brain connections for computational decision making and motor recovery [6, 13]. Given any two generalized smooth time series (mean and variance do not vary over time) X and Y, the Granger causality between them can be defined by an autoregressive model [14]:

$$X_t = \sum_{k=1}^{p} A_k Y_{t-k} + \sum_{k=1}^{p} B_k X_{t-k} + \xi_{xt} \quad (1)$$

$$Y_t = \sum_{k=1}^{p} A_k X_{t-k} + \sum_{k=1}^{p} B_k Y_{t-k} + \xi_{yt} \quad (2)$$

where, p is the order of the model, A_k is the Granger causality coefficient, B_k is the autoregressive coefficient, ξ_{xt} and ξ_{yt} is the prediction error. If the Granger causality coefficient A_k is 0, it means that the past of X_t has no help in the prediction of Y_t, that is, there is no causal relationship between X_t and Y_t. If the time series Y_t has a Granger causal effect on X_t, then its measure is calculated as [12]:

$$F_{Y \to X} = \ln \frac{\sum_{X|X^-}}{\sum_{X|X^-,Y^-}} \quad (3)$$

In the formula, X^- and Y^- represent the information of the previous moment of X and Y respectively. If X and Y are independent of each other, then $F_{X \to Y} = 0$. If has a Granger causal effect on X, then $F_{X \to Y} > 0$. Conversely, the Granger causality measure of X to Y can be calculated in the same way.

Conditional Granger causality, also known as multivariate Granger causality, is an improvement on Granger causality analysis and is more applicable when calculating the causal relationship of multiple time series. Suppose that for three variables X, Y and Z, in condition Z, the effect of Y on X can be defined as:

$$F_{Y \to X|Z} = \ln \frac{\sum_{X|X^-,Z^-}}{\sum_{X|X^-,Y^-,Z^-}} \quad (4)$$

If Y has no Granger causal effect on X, but an indirect effect through the condition Z, then $\sum_{X|X^-,Z^-} = \sum_{X|X^-,Y^-,Z^-}$, then $F_{Y \to X|Z} = 0$, that is, in the absence of Z, X and Y are independent of each other. If Y has a Granger causal effect on X under the conditional influence of Z, then $\sum_{X|X^-,Z^-} > \sum_{X|X^-,Y^-,Z^-}$, then $F_{Y \to X|Z} > 0$. From this, it can be judged whether there is a direct causal relationship between the two variables, and the influence of other indirect causal relationships can be excluded. In the actual EEG data processing process, other time series other than the causal relationship to be examined are usually used as condition sets. As condition sets, the situation of multiple variables can be regarded as three-variable conditional Granger causality.

3 Data Collection and Processing

3.1 Data Collection

The equipment selected for the experiment was an EGI EEG acquisition device. 11 healthy subjects, including 7 males and 4 females, with a mean age of 24.74 years ± 0.88

(standard deviation of the mean), normal or corrected eye vision, and all right-handed, were recruited for this study. All subjects had no history of psychiatric disorders and had no prior experience with either motor execution or motor imagery EEG experiments prior to this experimental study, and were fully informed of the procedure and voluntarily signed an informed consent form prior to the experiment.

Experimental paradigm design: In the experiment, the subjects were asked to sit in front of a computer screen at a distance of about 50 cm from their eyes and follow the instructions on the screen. In each trial, the finger motor execution and finger motor imagination were arranged within 15 s period, with each 15 s task consisting of a 3 s video of the finger gesture being played, a 3 s finger movement execution, a 3 s movement imagery, and three 2 s cues, which were played in sequence before the three tasks were performed. During which each subject had to perform the corresponding finger mortor execution and finger motor imagination according to the video content, and a total of 60 trials had to be completed, as shown in Fig. 1.

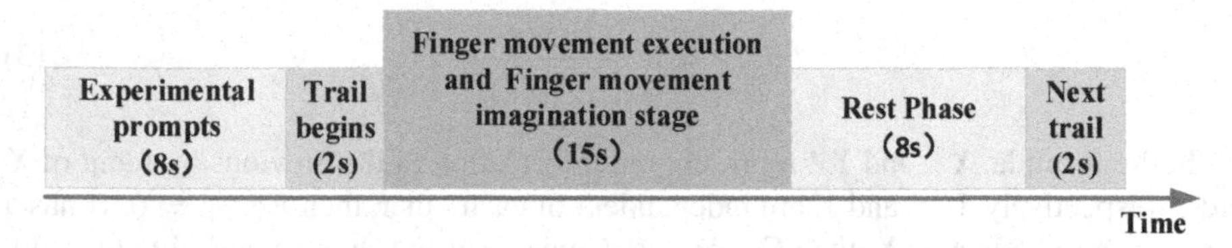

Fig.1. Basic flow chart of the experiment

3.2 Data Processing

Experimental data pre-processing: Firstly, the acquired EEG data were downsampled, and then the large amount of noise and artifacts interfered in the EEG signal were processed by using Butterworth filter and ICA independent component analysis, and then the EEG signal was further filtered by Laplace algorithm to improve the signal-to-noise ratio and obtain high quality signal.

Taking the experimental data collected by the EGI EEG acquisition equipment in this experiment as an example, the sampling frequency of the data is 1000 Hz. In order to reduce the computer memory and speed up the data operation, the original data is downsampled with a sampling factor of 5, and the signal frequency is 1000 Hz is reduced to 200 Hz. Figure 2 shows a comparison of the original signal and the downsampled signal.

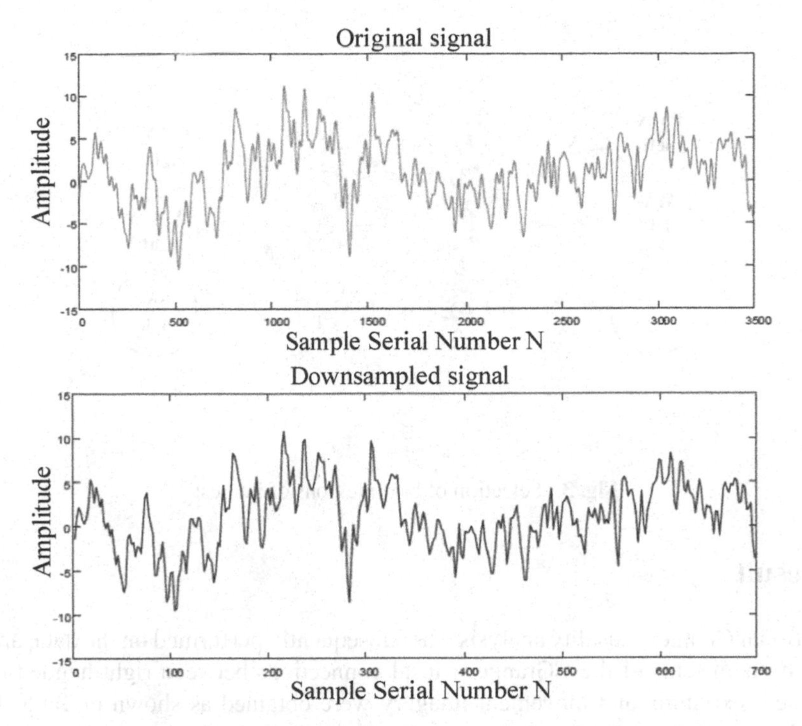

Fig. 2. EEG signal downsampling before and after signal comparison

Before conducting the Conditional Granger causality analysis, the definition of brain network nodes, that is, the definition of the region Of Interest (ROI), is firstly performed. Here, the left premotor area (PMA-left), the right premotor area (PMA-right), the left primary motor area (MA-left), the right primary motor area (MA-right), the left primary sensory area (SA-left) and the right primary sensory area (SA-right) are defined in the finger motor execution and motor imagination process. (MA-right), right primary motor area (MA-right), left primary sensory area (SA-left) and right primary sensory area (SA-right), a total of six regions, each containing four electrodes, as shown in Fig. 3.

After EEG signal preprocessing, the effective number of trails per subject per electrode is about 50 – 60, so that each region of interest contains at least 200 trails, and considering the complexity of calculation, 100 trails are randomly selected here as the EEG data signal of the region. And before the Granger causality calculation, the EEG data were subjected to the Durbin-Watson test, the purpose of which was to determine whether the residuals after the completion of the test regression obeyed the normal distribution, and if not, the basis of solving Granger causality was not satisfied. After calculation, the residuals of all subjects in this experiment satisfied the Durbin-Watson test criterion ($p > 0.60$), and the consistency tests all remained above 80%, and the excellent part of the TRAIL consistency test could reach 92%, which proved that the majority of the data were involved in the construction of the multiple autoregressive model.

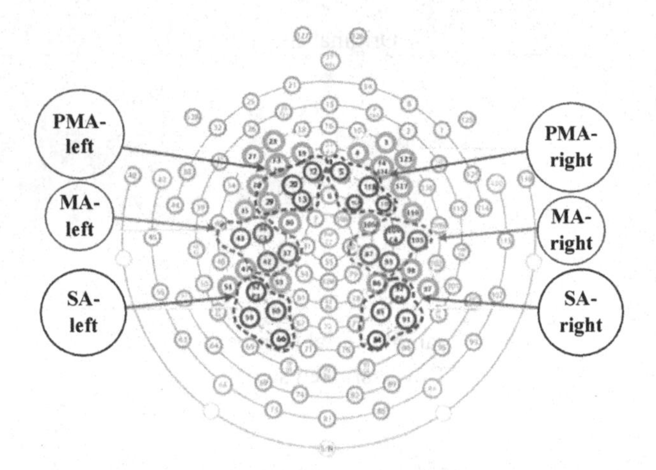

Fig. 3. Selection of brain regions of interest

4 Result

Conditional Granger causality analysis was subsequently performed on the data, and the strength coefficients of the CGranger causal connections between right-handed finger movement execution and movement imagery were obtained as shown in Table 1 and Table 2.

Table 1 and Table 2 show the conditional Granger causality coefficients for right-handed finger motor execution and motor imagery, respectively. p-value is a criterion to control for error rate through error correction rate (FDR), and its significance is the expected value of the number of false rejections (rejection of the true (original) hypothesis) as a proportion of the number of all rejected original hypotheses. The symbol #indicates a statistically significant probability value of $p < 0.05$, i.e. a statistically significant difference in the relative error values; and the symbol ※indicates a statistically significant threshold value of $p < 0.01$. i.e. an extremely significant statistical difference in the relative error values. The threshold value of $p < 0.05$ corrected by FDR in right-hand finger motor execution is 0.0201, and the thresh-old value of $p < 0.05$ corrected by FDR is 0.0400, Granger causality is significant when the threshold is greater than 0.0201 and extremely significant when the threshold is greater than 0.0400; the threshold value of $p < 0.05$ corrected by FDR in right-hand finger motor imagery is 0.0206 and $p <$ 0.01 corrected by FDR had a threshold of 0.0394, Similarly when the threshold value is greater than 0.0206 it means that the Granger causality between the two is significant and when the threshold value is greater than 0.0394 it means that the Granger causality between the two is extremely significant.

According to Table 1 and Table 2, the CGranger causality brain network diagrams of right-handed finger motor execution and motor imagery can be drawn. As shown in Fig. 3, the left subplot is the brain network diagram of right hand finger motor execution and the right subplot is the brain network diagram of right hand finger motor imagination, and the different thickness of arrows in the diagram represents different statistical significance.

Table1. CGranger causality coefficient for right-handed finger motor execution

To / From	PMA-left	MA-left	SA-left	PMA-right	SA-right
PMA-left	...	0.04507 ※	0.04856 ※	0.03026 ※	0.01949
MA-left	0.04233 ※	...	0.02135 #	0.01931	0.01944
SA-left	0.02145 #	0.04255 ※	...	0.01384	0.02176 #
PMA-right	0.02561 #	0.04575 ※	0.01486	...	0.02181 #
SA-right	0.01595	0.04836 ※	0.04499 ※	0.01436	...

#: p **<0.05 significance threshold of 0.0201;** ※: p **<0.01 significance threshold of 0.0400**

Table2. CGranger causality coefficient for right-handed finger motor imagery

To / From	PMA-left	MA-left	SA-left	PMA-right	SA-right
PMA-left	...	0.04354 ※	0.03673 #	0.02026 #	0.01840
MA-left	0.02499 #	...	0.02477 #	0.01074	0.01924
SA-left	0.02939 #	0.04567 ※	...	0.01505	0.04137 ※
PMA-right	0.01972	0.04352 ※	0.01705	...	0.03464 #
SA-right	0.01439	0.06477 ※	0.02593 #	0.01445	...

#: p **<0.05 significance threshold of 0.0206;** ※: p **<0.01 significance threshold of 0.0394**

From the Fig. 4, it can be seen that during the motor execution of the right hand finger, there are causally connected loops between the left MA, the left PMA, and the left SA of the contralateral cerebral hemisphere corresponding to the motor side, not only in the motor execution, and there are connected loops in the left and right PMA, and the left and right SA; while in the motor imagery, there are connected loops in the left and right SA, and there are connected loops from the left PMA to the In the motor imagination, there is a connection loop between the left and right SAs and a conditional Granger causal connection from the left PMA to the right PMA. Finally, there were conditional Granger causal connections from the right PMA to the right SA in the brain hemisphere ipsilateral to the motor hand. In summary, in the right-hand experiment, finger movement execution was associated with a valid connectivity circuit between the left PMA and the left MA, the left MA and the left SA, and the left SA and the right SA.

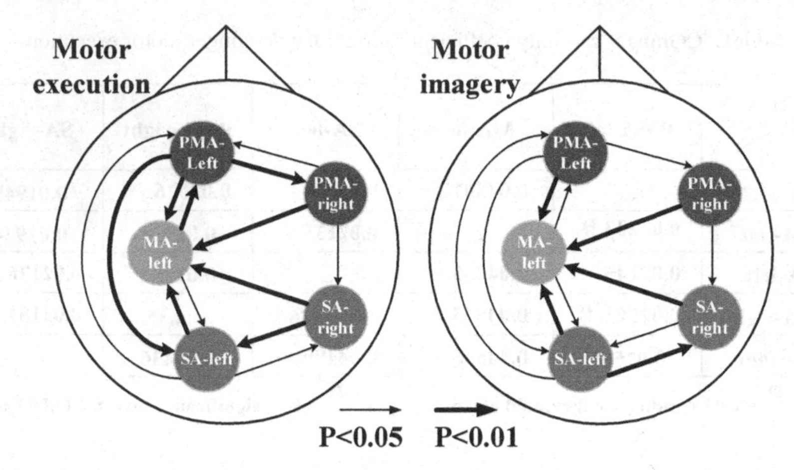

Fig. 4. CGranger causal brain network diagram of right-handed finger motor execution and motor imagery

5 Discussion

By analyzing the results of the CGranger causal brain net-work relationship between finger motor execution and motor imagery, we can see that some connecting parts of the CGranger causal brain network overlap during finger motor execution and motor imagery. These results demonstrate that motor imagery has physiological properties similar to those of motor execution. It is confirmed that the activated brain regions overlap during motor execution and motor imagery and that the interconnections between these regions are similar [10].

MA plays an important role in motor dynamics and motor control dynamics, such as finger movements and limb motor functions [15, 16], while PMA is involved in motor selection, motor planning, and motor preparation of the finger, such as the selection of motor procedures based on sensory information, or decision processes that rely on relevant knowledge learned earlier [16, 21]. In particular, PMA plays a very important role in motor response [17]. In experiments performed with visual stimuli cuing movement, visual information affecting movement is first processed by the associated sensory cortex and then is transmitted to non-primary motor areas [18]. Finally the combined information of somatosensory and visual information transmitted is received by the premotor area and guides movement based on sensory information [17].

In the study we found that the CGranger causal coefficient values from the PMA to the MA were somewhat larger than the values from the MA to the PMA. This is similar to the results found by previous researchers using TMS technique, indicating that neuromodulatory areas occurring after stimulation of the PMA are somewhat larger than those occurring after stimulation of the MA [19], revealing that the connections between the premotor area and bilateral primary motor areas are closely related [20–22].

In summary, this paper used CGranger causality analysis to analyse and compare the CGranger causal connectivity and connection strength between the most representative motor-activated brain regions: left and right premotor areas, contralateral primary motor areas and left and right primary sensory areas. The results showed that the brain network map of finger motor execution had more effective connections than that of finger motor imagery, indicating that the information transfer in motor imagery brain areas was weaker; in addition, effective connectivity circuits were found between left PMA and left MA, left MA and left SA, and left SA and right SA in both finger motor execution and motor imagery. The most important connections in motor function during finger motor execution and motor imagery were from the premotor area to the primary motor area. These findings may serve as valid features for motor function assessment and provide a reference for future clinical guidance of motor rehabilitation evaluation in patients with motor impairment after stroke in healthy populations.

References

1. Boe, S., Gionfriddo, A., Kraeutner, S., Tremblay, A., Bardouille, T.: Laterality of brain activity during motor imagery is modulated by the provision of source level neurofeedback. In: NeuroImage, 2014, Conference 2016. LNCS, vol. 9999, pp. 1–13. Springer, Heidelberg (2016)
2. He, L., Hu, D., Meng, W., Ying, W., Deneen, K.M.V., Zhou, M.C.: Common bayesian network for classification of EEG-based multiclass motor imagery BCI. IEEE Trans. Syst. Man Cybern. Syst. (2017)
3. Arvaneh, M., et al.: Facilitating motor imagery-based brain–computer interface for stroke patients using passive movement. Neural Comput. Appl. **28**(11), 3259–3272 (2016). https://doi.org/10.1007/s00521-016-2234-7
4. Fingelkurts, A.A., Kahkonen, S.: Functional connectivity in the brain is it an elusive concept. Neurosci. Biobehav. Rev. **28**(8), 827–836 (2005)
5. Petit, L., Orssaud, C., Tzourio, N., Mazoyer, B., Berthoz, A.: Do Executed, Imagined and Suppressed Saccadic Eye Movements Share the Same Neuronal Mechanisms in Healthy Human ? Springer, Netherlands (1996)
6. Chen, H., Yang, Q., Liao, W., et al.: Evaluation of the effective connectivity of supplementary motor areas during motor imagery using Granger causality mapping. Neuroimage **47**(4), 1844–1853 (2009)
7. Friston, K.J., Frith, C.D., Liddle, P.F., Frackowiak, R.S.J.: Functional connectivity: the principal-component analysis of large (PET) data sets. J. Cereb. Blood Flow Metab. **13**(1), 5–14 (1993)
8. Friston, K.J., Frith, C.D., Frackowiak, R.S.J.: Time-dependent changes in effective connectivity measured with PET. Hum. Brain Mapp. **1**, 69–79 (1993)
9. Lacourse, M.G., Orr, E.L.R., Cramer, S.C., Cohen, M.J.: Brain activation during execution and motor imagery of novel and skilled sequential hand movements. Neuroimage **27**(3), 505–519 (2005)
10. Solodkin, A., Hlustik, P., Chen, E.E., Small, S.L.: Fine modulation in network activation during motor execution and motor imagery. Cereb. Cortex **14**, 1246–1255 (2004)
11. Wang, S., Zhan, Y., Zhang, Y., et al.: Abnormal long- and short-range functional connectivity in adolescentonset schizophrenia patients: a resting-state fMRI study. Progress in Neuro-Psychopharmacol. Biol. Psychiatry **81**, 445–451 (2018)
12. John, G.: Measurement of linear dependence and feedback between multiple time series. J. Am. Stat. Assoc. **77**(378), 304–313 (1982)
13. Friston, K., Moran, R., Seth, A.K.: Analysing connectivity with Granger causality and dynamic causal modelling. Curr. Opin. Neurobiol. **23**(2), 172–178 (2013)
14. Ding, M., Chen, Y., Bressler, S.L.: Granger Causality: Basic Theory and Application to Neuroscience. John Wiley & Sons, Ltd (2006)

15. Arvaneh, M., et al.: Facilitating motor imagery based brain computer interface for stroke patients using passive movement. Neural Comput. Appl. (2016)s
16. Rushworth, M.F.S., Johansen-Berg, H., GöBel, S.M., Devlin, J.T.: The left parietal and premotor cortices: motor attention and selection. Neuoimage **20**, S89–S100 (2003)
17. Chouinard, P.A.: The primary motor and premotor areas of the human cerebral cortex. Neuroscientist **12**(2), 143–152 (2006)
18. Ehrsson, H.: Imagery of voluntary movement of fingers, toes, and tongue activates corresponding body part specific motor representations. J. Neurophysiol. **90**(5), 3304–3316 (2003)
19. Reis, J., Swayne, O.B., Vandermeeren, Y., Camus, M., Cohen, L.G.: Contribution of transcranial magnetic stimulation to the understanding of cortical mechanisms involved in motor control. J. Physiol. **586**(2), 325–351 (2010)
20. Hammer, J., et al.: Predominance of movement speed over direction in neuronal population signals of motor cortex: intracranial EEG data and a simple explanatory model. Cereb. Cortex **26**, 2863–2881 (2016)
21. Wise, S.P., Boussaoud, D., Johnson, P.B., et al.: Premotor and parietal cortex: corticocortical connectivity and combinatorial computations. Annu. Rev. Neurosci. **20**(20), 25 (1997)
22. Dum, R.P.: Frontal lobe inputs to the digit representations of the motor areas on the lateral surface of the hemisphere. J. Neurosci. **25**(6), 1375–1386 (2005)

A Hybrid Brain-Computer Interface for Smart Car Control

Nianming Ban, Chao Qu(✉), Daqin Feng, and Jiahui Pan

School of Software, South China Normal University, Guangzhou 510631, China
quchao@m.scnu.edu.cn

Abstract. Brain-computer interface (BCI) systems are often used to convert signals from brain activities into control commands through external devices. There are few studies on controlling a car by multi-modality due to its difficulty in the current research. This paper proposes a hybrid BCI control system based on electroencephalography (EEG), electrooculography (EOG), and gyroscope signals to address this challenge. The user can control the start, stop, turn left, turn right, acceleration and deceleration of the smart car by this system. The user controls the start and stop by double blinking, acceleration and deceleration by concentrating and distracting, turning left and right by the head rotation. To evaluate the performance of this BCI system, we invited twelve subjects to conduct two online experiments to control the car on a runway to test the above functions. The experimental results showed that the hybrid BCI system achieved an average accuracy of 97.65%, an average information translate rate (ITR) of 43.50 bit/min, and an average false positive rate (FPR) of 0.70 event/min, thus demonstrating the effectiveness of our proposed system.

Keywords: Brain-Computer Interface (BCI) · Smart car · Multimodal control

1 Introduction

In traditional control systems, people rely on traditional keyboards and joysticks to control external devices, but this approach is challenging to achieve for people with severe motor disabilities. Thus, a new bioelectric signal-based human-machine interface has emerged, which can recognize different control commands directly from the user's bioelectric signal and can be used to help people with severe disabilities, such as amyotrophic lateral sclerosis (ALS), spinal cord injury (SCI), and stroke, to communicate with the outside environment again.

In recent years, EOG and EEG based hybrid BCI systems have been widely used in more complex settings [1], the using scenario has shifted from the laboratory to our daily life [2,3], and have proven to be noninvasive, inexpensive, convenient, and efficient. Among the EEG-based BCI systems used are motor imagery (MI) [4], P300 potentials [5], steady-state visual evoked potentials (SSVEPs) [6],

X. Ying (Ed.): HBAI 2022, CCIS 1692, pp. 135–147, 2023.
https://doi.org/10.1007/978-981-19-8222-4_12

motor onset visual evoked potentials (MVEPs) [7], and their mixed modalities [8,9].

In [10], Long et al. proposed an SSVEP-based asynchronous paradigm to control the car, and they used six stimulus frequencies to generate a multitask vehicle control strategy including turn left and right signals, wipers, horn, doors, and hazard lights. Four subjects participated in an online car control experiment, an average accuracy of 88.43% was achieved. In [11], Wang et al. proposed multiple patterns of motor imagery BCI to control the turn left and right and acceleration/deceleration of the car through the motor imagery of left-hand-right motor imagery, foot motor imagery and two-hand motor imagery, the metric of path length optimality ratio is 1.23 and the time optimality ratio is 1.28.

Also, EOG signal-based control systems have been widely used to control external devices because eye movements, such as blinking and eye rotation, can produce significant signal changes for feature extraction. In [12], Huang et al. proposed an EOG signal-based wheelchair control system, which was designed with an interface containing 13 blinking buttons, each corresponding to a command blinked one by one in a predefined order, and the subject performed blinking simultaneously to select the command according to the blinking buttons, the system achieved an accuracy rate of 96.7%. In [13], Olesen et al. proposed a hybrid EEG-EOG BCI system to control wheelchair by combining the MI-EEG signal and EOG signal to control the movements such as turn left and right and forward-backward of the vehicle with an accuracy of 87.3%.

However, there are some defects need to be improved in the system described above:

1. Multimodal control: The previous BCI systems, whether based on SSVEP, P300 or MI, control the device through only one modal. The system needs to process one signal before processing the next; it cannot turn left or turn right while controlling the car acceleration. It leads to the fact that unimodal systems are not suitable for working in complex scenarios. Therefore, we propose a hybrid BCI control system to the car in three modals: direction control (turn left/right), speed control (acceleration/deceleration) and start/stop control. All three signals extracted operate independently and do not affect each other. In addition, multimodal control solves the problem of the insufficient number of commands. In previous studies, the number of control commands was also relatively limited. For example, in the MI-based BCI system, the number of commands generated through the brain is less and takes much time to train. In EOG-based control systems, it is also not possible to generate more control commands because the actions that the eye can generate are limited in blinking and rotation [14]. Therefore, the multimodal control system which fuses the EOG, EEG and gyroscope can provide multiple control commands for complex operations.

2. Convenient and easy-to-use system: In previous BCI systems, when selecting the acquisition device for the signal, researchers have proposed the use of an EEG cap for EEG signal acquisition. Although this multi-channel acquisition method can increase the robustness of the acquired signal, the use of an EEG cap is inconvenient in daily life, which requires a cumbersome wearing method and

the application of the conductive gel. We chose to use HNNK's dry electrode portable brain loop, which can be worn directly by the user, increasing user convenience. In addition, previous BCI systems have been less than satisfactory in use, taking much time for training in MI-based BCI systems and fatiguing subjects in SSVEP-based or P300-based BCI systems. The system proposed in this paper allows subjects to become skilled with it quickly and without fatigue, which makes it more suitable for daily wear by people with disabilities who need a BCI system.

Therefore, in our study, a hybrid BCI system combining EOG, EEG and gyroscope is proposed to control the multi-dimensional car functions of start, stop, left turn, right turn, acceleration, and deceleration. To verify the effectiveness of this BCI system, twelve subjects aged 22–28 were asked to control the car to perform six operations, including start, stop, acceleration, deceleration, turn left and turn right, and complete the specified tasks on a set track. The results showed that the BCI system was 97.65% accurate, with an average ITR of 43.50 bit/min and an average FPR of 0.70. The subjects were able to use the BCI system to control the car well, thus confirming the efficiency and effectiveness of the system, on the other hand, suggesting that the system may improve self-care for people with severe motor disabilities.

2 Materials and Methods

2.1 System Overview

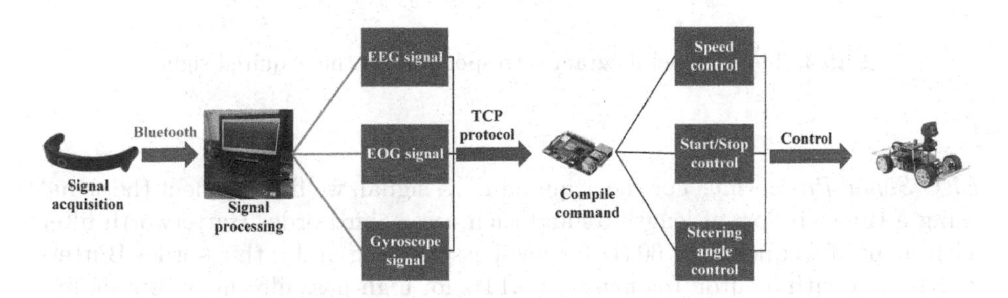

Fig. 1. Overall flow chart of multimodal control system

This BCI system combines EEG signal, EOG signal and gyroscope signal to achieve multimodal asynchronous control of the car, as shown in Fig. 1 The system consists of three main subsystems: signal acquisition, signal processing, and the intelligent car's control system. Firstly, the original EEG signal, EOG signal and gyroscope angle change signal are collected using HNNK equipment and transmitted to three different data processing programs in the computer terminal to pre-process and classify. The processed signals are then transmitted to the Raspberry Pi via TCP protocol. Finally, the car is driven by the Raspberry Pi. The car uses a Raspberry Pi 4B motherboard and an L298N motor.

2.2 Signal Acquisition

In this study, we used a signal acquisition equipment from the HNNK company to acquire the EOG signal, EEG signal and gyroscopic signal, which were acquired through three dry electrodes ("CH1", "COM", and "COMLEG"), the impedance between the electrodes and the skin is below 5 kΩ, the sampling rate of the device is 125 Hz, and it has a 50 Hz industrial frequency filtering. There is a gyroscope at the center of the device to acquire the user's head rotation information. As shown in Fig. 2, three electrodes are located at the two temples and forehead, with the "CH1" electrode used to acquire the EOG signal.

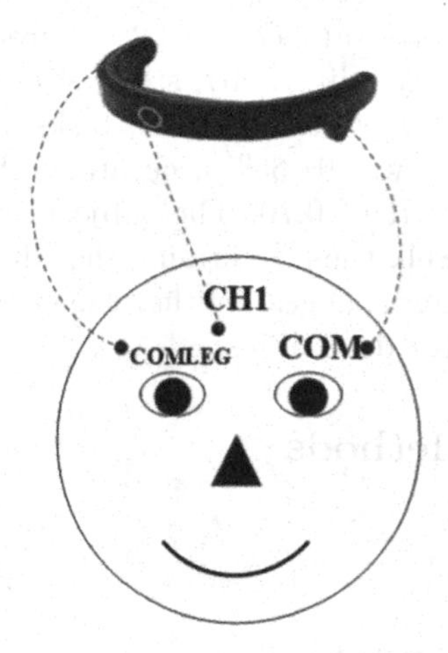

Fig. 2. The channel diagram corresponding to the acquired signal

EEG Signal Processing. For the acquired EEG signal, we first segment the signal using a time window of length 5 s and then use a third-order Butterworth filter with a cutoff frequency of 60 Hz for low-pass filtering and a third-order Butterworth filter with a cutoff frequency of 0.1 Hz for high-pass filtering to obtain five different frequency bands of waveforms, namely δ (1 $\sim$ 3 Hz), θ (4 $\sim$ 7 Hz), α (8 $\sim$ 13 Hz), β (14 $\sim$ 30 Hz) and γ (31 $\sim$ 48 Hz). Among them, the attention-related signals are α, β, θ and γ signals [3,15], respectively.

Then, we use the Welch algorithm (Eq. (1)) to calculate the power spectral density of each frequency band for feature extraction.

$$PSD_i(\mathit{freq}) = \frac{1}{MU}\left|\sum_{n=0}^{M-1} x_i(n)w(n)e^{-j2\pi \mathit{freq}_i}\right|^2 \tag{1}$$

where *w(n)* is the Hamming window function, *M* is the length of the segment, i denotes the i-th segment of the EEG signal, and *U* is the calculated power of the window function. The formula is as follows:

$$U = \frac{1}{M} \sum_{n=0}^{M-1} w^2(n) \tag{2}$$

We set P_{freq} to be the energy value corresponding to each segment of *freq*, and then calculate the energy values corresponding to the five eigenfrequencies by the following equation:

$$P_{freq} = \frac{1}{4} \sum_{i=0}^{3} PSD_i(freq) \tag{3}$$

$$E_\delta = \sum_{freq=1}^{3} P_{freq} \tag{4}$$

$$E_\theta = \sum_{freq=4}^{7} P_{freq} \tag{5}$$

$$E_\alpha = \sum_{freq=8}^{13} P_{freq} \tag{6}$$

$$E_\beta = \sum_{freq=14}^{30} P_{freq} \tag{7}$$

$$E_\gamma = \sum_{freq=31}^{48} P_{freq} \tag{8}$$

The support vector machine (SVM) with linear function as kernel function is immediately used to classify and identify the five features. Among them, we collect the attention samples for training and testing through offline experiments, and find the best parameters of the model for tuning the classifier through cross-validation.

EOG Signal Processing. For the acquired EOG signal, we filtered it first using a band-pass filter of 1 Hz–10 Hz to remove the effects caused by high-frequency noise and baseline drift [16], then we used a projection curve-based lookup algorithm to find the wave crest. For the acquired sequence of sampling points, we assume that the projection curve can be expressed as:

$$V = [v_1, v_2, \ldots, v_n] \tag{9}$$

Then calculate the first-order difference vector *Diff* of *V*.

$$Diff_v(i) = V(i+1) - V(i) \tag{10}$$

The difference vector is then subjected to a sign-taking function operation and recorded as $T = sign(Diff_v)$.

$$sign(x) = \begin{cases} 1 & if\ f > 0 \\ 0 & if\ f = 0 \\ -1 & if\ f < 0 \end{cases} \tag{11}$$

Then traversing the T from the tail as follows:

$$T(i) = \begin{cases} -1 & if\ \ T(i) = 0 \& T(i+1) < 0 \\ 1 & if\ \ T(i) = 0 \& T(i+1) \geq 0 \end{cases} \tag{12}$$

Then the first-order difference operation is performed on the obtained T to finally obtain the R. If $R(i) = -2$, then $i + 1$ is a peak bit of the projection V, corresponding to the peak $V(i+1)$. The final waveform we detected is shown in Fig. 3.

$$R = Diff(T) \tag{13}$$

The two wave crests correspond to moments t_{peak1}, t_{peak2}, The voltage value corresponding to the moment t of the sampling point is Vt. The time interval G between the two blink wave crests and the generated energy value E are calculated by the following equation:

$$G = t_{peak2} - t_{peak1} \tag{14}$$

$$E = \sum_{t=t_{peak1}}^{t=t_{peak2}} V_t^2 \tag{15}$$

Then compare the calculated time interval and energy values with the set thresholds to determine whether a double blink signal has been generated, a double blink is produced if the following conditions are met:

$$\begin{cases} G_{min} \leq G \leq G_{max} \\ E_{min} \leq E \leq E_{max} \end{cases} \tag{16}$$

Because of individual differences, G_{min}, G_{max}, E_{min} and E_{max} are different for each person. We designed a user's blink calibration procedure for generating each user's own thresholds. Specifically, prior to the start of the experiment, users were asked to perform a double blink operation every 3 s for a total of 5 times. The acquired signals were sliced using a 3 s time window, and then each slice was processed to obtain the feature values G and E by the signal processing and feature extraction methods mentioned above. The maximum and minimum values of the time interval G were recorded as G_{max} and G_{min}, and the maximum and minimum values of the energy value E were recorded as E_{max} and E_{min}.

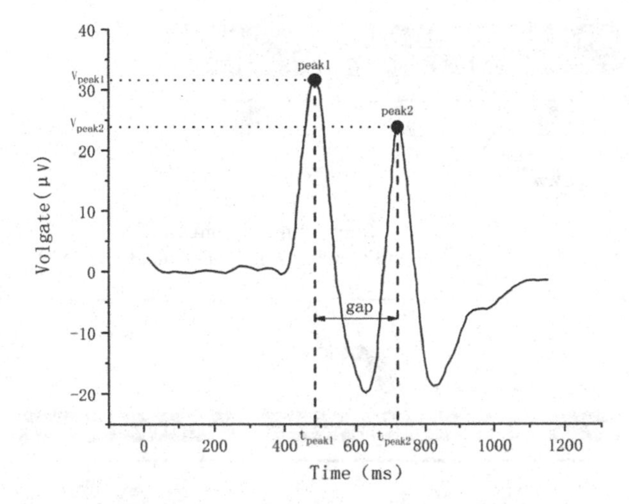

Fig. 3. Voltage changes produced by double blinking

Gyroscope Signal Processing. The rotation data of the gyroscope comes from the built-in sensor module, which detects the transformation of two-dimensional coordinates. We only perform feature extraction on the rotation data in the horizontal direction. The program sets an initial coordinate when the user wears it, and the magnitude of deflection is obtained by calculating the difference between the current coordinate and the initial coordinate when the user's head is turned left and right. The calculation formula is shown below:

$$\Delta L = \frac{\Delta\ yaw}{pi \times 180} \times (1 + s) \tag{17}$$

$$L_{new} = L_{initial} + \Delta L \tag{18}$$

where Δyaw is the change in horizontal direction detected by the sensor. s is a factor to adjust the sensitivity of the cursor ranging from -0.5 to 0.5. $L_{initial}$ is the initial position of the program record. L_{new} is the deflection coordinate calculated by the program.

3 Experiments and Results

To validate the effectiveness of this BCI system and assess the performance of the control car, we invited twelve healthy subjects (six males and six females, mean age 24 years in aged 22–28 years), all with normal vision and without any cognitive impairment, to participate in two online experiments. Each subject read and completed an informed consent form prior to conducting the experiments. The performance indices used in this study are listed below:

1. Accuracy (ACC): the possibility of a correct control command.

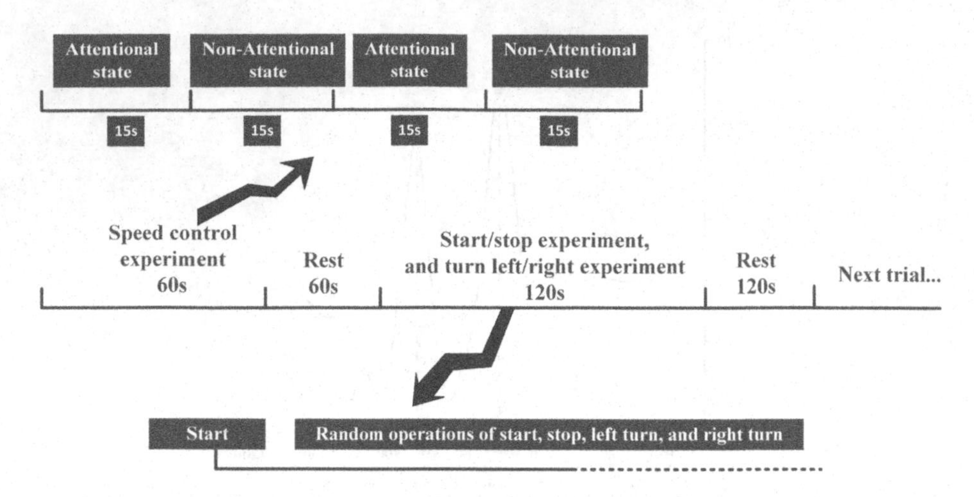

Fig. 4. Flow chart of Experiment I

2. False positive rate (FPR): false commands generated per minute during idle time.
3. Information transfer rate (ITR): the number of bits of information transferred per minute.

3.1 Experiment I: Single-Mode Control

To verify the effectiveness of single-modal control on the system and improve subjects' proficiency in controlling the system in preparation for Experiment II, we first conducted a single-modal control experiment, which was divided into three sub-experiments. In the first experiment, subjects were asked to control attentional and non-attentional states, which were used to generate commands to control the speed of the car. Specifically, after hearing the start command, subjects were asked to focus their attention through meditation, such as mental arithmetic or imagining an event, and the meditation time was set to 15 s. Immediately afterward, subjects were asked to be a non-attentional state, and the non-attentional state was also set to 15 s. The attentional and non-attentional states were performed again immediately.

After a 1-min rest, the start/stop experiment and the turn left/right experiment were conducted simultaneously, and the results were recorded separately. In these two experiments, the subjects controlled the start and stop by double blinking, and then the head rotation controlled the turn left and right of the car. Specifically, after hearing the start command, the subject controlled the start of the car by double blinking, followed by random turn left, turn right, stop and start commands issued, one every 3 s on average. The subject completed the corresponding operation after hearing the command, and the process lasted about 2 min.

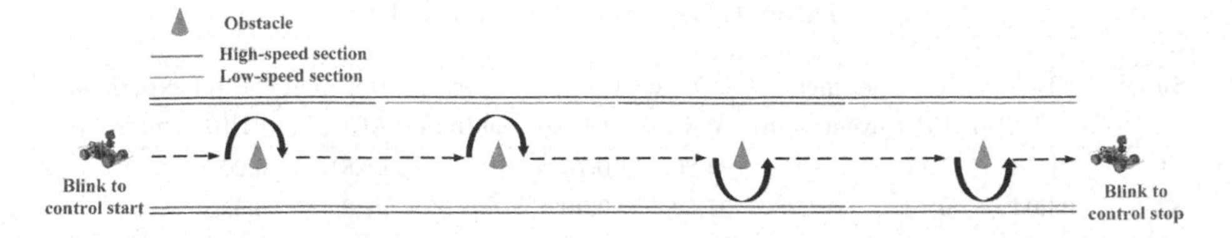

Fig. 5. Runway designed for multimodal control of intelligent car experiment

The above experimental process was recorded as one trail, and the next trail of experiments continued after two minutes. A total of 5 trails of experiments. Experiment I flow chart is shown in Fig. 4. The effectiveness of individual modal control is evaluated by calculating ACC and FPR from the experimental results.

3.2 Experiment II: Multimodal Car Control

In this experiment, we tested whether the user could use the system to control the car through three modalities simultaneously. And we designed a runway, as shown in Fig. 5, with a runway length of 8 m. The runway contains four sections, divided into two kinds of sections: high-speed section and low-speed section (the green section is the high-speed section, and the orange section is the low-speed section). The maximum speed of the car during the experiment was set to 40 m/min, the minimum speed was set to 10 m/min. The steering angle was set from −70 to +70 (The direction of the car going straight is set to 0°). When the experiment started, the subjects controlled the start of the car by double blinking at the starting point. Then, in the high-speed section, they were asked to accelerate the car by concentration. In the low-speed section, they were asked to decelerate the car by distraction. An obstacle was set up in the center of each section. The subjects were asked to control the car to turn left around the obstacle in the first and second sections, and to turn right around the obstacle in the third and fourth sections. Finally, reaching the endpoint and stopping the car with a double blink. When driving, the subjects can control the car to accelerate and decelerate while turning left and right, and can control the car to start and stop at any time by blinking. This confirms that our system can be independently controlled by the three signals without affecting each other. Each subject completed five experiments, also to prove the efficiency of this system, the completion time, ACC, and ITR were used to evaluate this hybrid BCI system.

Table 1. Results of Experiment I

Subject	EEG control experiment		EOG control experiment		Rotation control experiment	
	ACC(%)	FPR(event/min)	ACC(%)	FPR(event/min)	ACC(%)	FPR(event/min)
S1	91.86	1.81	94.20	0.75	98.81	0.27
S2	93.04	0.78	97.53	0.30	97.89	0.34
S3	96.48	0.92	97.60	0.30	96.48	0.74
S4	95.33	0.89	96.45	0.60	95.60	0.97
S5	93.79	1.31	98.40	0.20	99.04	0.29
S6	94.58	1.13	96.00	0.40	98.47	0.39
S7	95.23	0.66	97.83	0.50	97.21	0.42
S8	94.27	1.45	98.20	0.30	96.63	0.85
S9	98.17	0.46	96.50	0.45	97.85	0.72
S10	95.38	0.83	98.40	0.15	98.67	0.24
S11	93.41	1.63	92.83	0.50	96.45	0.82
S12	98.67	0.32	96.00	0.25	97.75	0.36

3.3 Results

All subjects completed the experiments as required, and the results of unimodal control for Experiment I are shown in Table 1. Table 2 shows the multimodal control results of Experiment II. All the data shown in the table are the calculated averages. The experimental results show that the hybrid BCI system we designed is highly accurate and efficient.

Table 2. Results of Experiment II

Subject	Time(s)	ACC(%)	FPR(event/min)	ITR(bit/min)
S1	21.53	94.86	1.20	45.12
S2	18.69	96.04	0.95	55.39
S3	28.56	97.48	0.89	37.56
S4	23.02	98.33	0.71	41.37
S5	31.23	96.79	0.88	35.79
S6	19.64	97.58	0.45	52.44
S7	27.67	98.23	0.31	48.62
S8	22.53	98.89	0.63	44.86
S9	21.32	97.67	0.69	40.45
S10	32.34	98.89	0.92	36.09
S11	18.94	98.64	0.46	41.37
S12	19.52	98.43	0.33	42.93

The experimental results show that in Experiment I, the average accuracy of the EEG control experiment, blink control experiment, and rotation control experiment reached 95.02%, 96.66% and 97.57%, respectively, and the average FPR reached 1.02 event/min, 0.39 event/min and 0.53 event/min. The average accuracy of Experiment II reached 97.65%, the average FPR reached 0.70 event/min, and the average ITR reached 43.50 bit/min. These two online experiments demonstrate the effectiveness of the system. The three signals can work independently without affecting each other, allowing the subjects to conveniently control the car.

As shown in Table 3, in other people's research, the brain patterns they use are only one or two, and we use three (EEG+EOG+Gyroscope), which gives us a maximum of six control dimensions of control, and other people's research has only four or five, which allows us to control the external devices more convenient. Not only that, we also achieved the highest accuracy rate of 97.65%. And the subjects can control the car proficiently to complete the task after training, the average completion time of experiment II is only 23.75 s, which shows that the system is easy to learn, with high accuracy and high practicality.

Table 3. Comparison of the results of this paper with those of studies [13,17–19]

Works	Brain patterns	Control objects	Control dimension	Accuracy(%)
Liu et al.	SSVEP	Trolley	4 (forward, backward, turn left, and turn right)	87.50
Olesen et al.	EEG+EOG	Wheelchair	4 (forward, reverse, turn left, and turn right)	87.30
Zhou et al.	MI	Virtual Car	4 (forward, reverse, turn left, and turn right)	73.66
Farmaki et al.	SSVEP	Trolley	5 (forward, backward, turn left, turn right, and stop)	81.00
Our work	**EEG+EOG+ Gyroscope**	Trolley	**6 (start, stop, turn left, turn right, acceleration, and deceleration)**	**97.65**

4 Conclusions

A key issue in the control system of hybrid BCI is how to provide a sufficient number of commands to the control device and improve the accuracy rate to meet the device to perform more complex tasks. This paper presents a hybrid BCI system that combines EEG, EOG, and gyroscope signals to control the car functions of start, stop, turn left, turn right, acceleration, and deceleration. Specifically, the user controls the acceleration and deceleration by EEG detection, implements the start and stop by double blinking, and turn left and right by head rotation. We conducted two online experiments, one using unimodal control of the car and the other using multimodal control of the car on the runway to complete the task. Both experiments confirmed that our method and system are effective. From further experimental results, it is clear that the proposed hybrid BCI system we used not only provides multiple independent signals for higher complexity control, but also improves the accuracy rate. However, this system still needs to be improved in BCI-based functions. For example, when using the EEG signal to control the car, the speed change of the car is not obvious. This requires an update to our algorithm for processing EEG signals. In the future, we plan to use a more advanced CNN network to process the EEG signal to achieve better classification results.

References

1. Li, Z., Zhang, S., Pan, J.: Advances in hybrid brain-computer interfaces: principles, design, and applications. Comput. Intell. Neurosci. **2019**, 1–9 (2019). https://doi.org/10.1155/2019/3807670
2. Wang, F., Li, X., Pan, J.: A human-machine interface based on an EOG and a gyroscope for humanoid robot control and its application to home services. J. Healthc. Eng. **2022**, 1–14 (2022). https://doi.org/10.1155/2022/1650387
3. Cai, X., Pan, J.: Toward a brain-computer interface- and internet of things-based smart ward collaborative system using hybrid signals. J. Healthc. Eng. **2022**, 1–13 (2022). https://doi.org/10.1155/2022/6894392
4. Pfurtscheller, G., da Silva, F.L.: Event-related EEG/MEG synchronization and desynchronization: basic principles. Clin. Neurophysiol. **110**(11), 1842–1857 (1999). https://doi.org/10.1016/s1388-2457(99)00141-8
5. Farwell, L., Donchin, E.: Talking off the top of your head: toward a mental prosthesis utilizing event-related brain potentials. Electroencephalogr. Clin. Neurophysiol. **70**(6), 510–523 (1988). https://doi.org/10.1016/0013-4694(88)90149-6
6. Zhang, Y., Zhou, G., Jin, J., Wang, X., Cichocki, A.: Frequency recognition in SSVEP-based BCI using multiset canonical correlation analysis. Int. J. Neural Syst. **24**(04), 1450013 (2014). https://doi.org/10.1142/s0129065714500130
7. Guo, F., Hong, B., Gao, X., Gao, S.: A brain–computer interface using motion-onset visual evoked potential. J. Neural Eng. **5**(4), 477–485 (2008). https://doi.org/10.1088/1741-2560/5/4/011
8. Li, Y., Pan, J., Wang, F., Yu, Z.: A hybrid BCI system combining p300 and SSVEP and its application to wheelchair control. IEEE Trans. Biomed. Eng. **60**(11), 3156–3166 (2013). https://doi.org/10.1109/tbme.2013.2270283
9. Ma, T., Li, H., Deng, L., Yang, H., Lv, X., Li, P., Li, F., Zhang, R., Liu, T., Yao, D., Xu, P.: The hybrid BCI system for movement control by combining motor imagery and moving onset visual evoked potential. J. Neural Eng. **14**(2), 026015 (2017). https://doi.org/10.1088/1741-2552/aa5d5f
10. Long, S., Zhou, Z., Yu, Y., Liu, Y., Zhang, N.: Research on vehicle control technology of brain-computer interface based on SSVEP. In: Su, R. (ed.) 2019 International Conference on Image and Video Processing, and Artificial Intelligence. SPIE (2019). https://doi.org/10.1117/12.2549159
11. Wang, H., Li, T., Bezerianos, A., Huang, H., He, Y., Chen, P.: The control of a virtual automatic car based on multiple patterns of motor imagery BCI. Med. Biol. Eng. Comput. **57**(1), 299–309 (2018). https://doi.org/10.1007/s11517-018-1883-3
12. Huang, Q., et al.: An EOG-based human–machine interface for wheelchair control. IEEE Trans. Biomed. Eng. **65**(9), 2023–2032 (2018). https://doi.org/10.1109/tbme.2017.2732479

13. Olesen, S.D.T., Das, R., Olsson, M.D., Khan, M.A., Puthusserypady, S.: Hybrid EEG-EOG-based BCI system for vehicle control. In: 2021 9th International Winter Conference on Brain-Computer Interface (BCI). IEEE (2021). https://doi.org/10.1109/bci51272.2021.9385300
14. Milanizadeh, S., Safaie, J.: EOG-based HCI system for quadcopter navigation. IEEE Trans. Instrum. Meas. **69**(11), 8992–8999 (2020). https://doi.org/10.1109/tim.2020.3001411
15. Liu, N.H., Chiang, C.Y., Chu, H.C.: Recognizing the degree of human attention using EEG signals from mobile sensors. Sensors **13**(8), 10273–10286 (2013). https://doi.org/10.3390/s130810273
16. Ma, J., Zhang, Y., Cichocki, A., Matsuno, F.: A novel EOG/EEG hybrid human-machine interface adopting eye movements and ERPs: application to robot control. IEEE Trans. Biomed. Eng. **62**(3), 876–889 (2015). https://doi.org/10.1109/tbme.2014.2369483
17. Liu, C., Xie, S., Xie, X., Duan, X., Wang, W., Obermayer, K.: Design of a video feedback SSVEP-BCI system for car control based on improved MUSIC method. In: 2018 6th International Conference on Brain-Computer Interface (BCI). IEEE (2018). https://doi.org/10.1109/iww-bci.2018.8311499
18. Zhou, Z., Gong, A., Qian, Q., Su, L., Zhao, L., Fu, Y.: A novel strategy for driving car brain–computer interfaces: discrimination of EEG-based visual-motor imagery. Transl. Neurosci. **12**(1), 482–493 (2021). https://doi.org/10.1515/tnsci-2020-0199
19. Farmaki, C., Krana, M., Pediaditis, M., Spanakis, E., Sakkalis, V.: Single-channel SSVEP-based BCI for robotic car navigation in real world conditions. In: 2019 IEEE 19th International Conference on Bioinformatics and Bioengineering (BIBE). IEEE (2019). https://doi.org/10.1109/bibe.2019.00120

A Spiking Neural Network for Brain-Computer Interface of Four Classes Motor Imagery

Yulin Li, Hui Shen(✉), and Dewen Hu

College of Intelligence Science, National University of Defense Technology, Changsha, China
shenhui@nudt.edu.cn

Abstract. Spiking neural networks (SNN) has the advantages of low power consumption and high efficiency in processing temporal information. However, due to the difficulty of network training, there exist few studies about the applications of SNN in brain-computer interface (BCI), especially in the four-classification task of motor imagery (MI). In this study, we develop a four-layer SNN structure to solve the MI four-classification problem. Firstly, an improved optimization algorithm for Ben's spiker algorithm (BSA) is presented to convert EEG signals into spike signals, which obtains about 50 times higher efficiency than the commonly used optimizing algorithms. Secondly, a SNN combined with spike long-short-time-memory (LSTM) module is proposed to perform four-classification tasks in MI. Finally, we introduce the channel-wise normalization strategy to facilitate the training of deeper layers. Our experiment on the publicly released dataset achieves the accuracy that is comparable to the previous work of one-Dimension convolution neural network (1D-CNN). Meanwhile, the number of parameters of proposed network is about 1/10 of that in 1D-CNN. This study reveals the great potential of the SNN in developing a low-power and wearable BCI system.

Keywords: Brain-computer interface · Motor imagery · Spiking neural networks

1 Introduction

Spiking neural networks (SNN), as a neural network inspired by brain science and with high biological reliability. Currently, SNN has been widely used to construct brain-like intelligence. Compared with artificial neural network (ANN), SNN has two unique advantages. Firstly, the neuron model used by SNN is more similar to biological brain neurons that has rich dynamic characteristics and inherent ability to process temporal information. Secondly, SNN accepts and fires spike trains, so that its computational process is sparse in the timing scale, leading to extremely low energy consumption and high response efficiency in the process of speculating and training [1, 2].

As neurons in SNN are driven by spikes to update the internal state, the input of SNN must be spike trains, so that spike coding strategies are indispensable for converting sample data into spike trains. Main encoding methods can be categorized into frequency-based and temporal-based methods. For electroencephalogram (EEG) signals used in this work, Ben's spiker algorithm (BSA) is an ideal choice, it is a temporal encoding method

X. Ying (Ed.): HBAI 2022, CCIS 1692, pp. 148–160, 2023.
https://doi.org/10.1007/978-981-19-8222-4_13

and usually be used to convert temporal data [3, 4]. As the convert effect of BSA is related to multiple parameters, previous works commonly use grid search or genetic algorithm to perform optimization. However, the optimization efficiency of those two methods is rather low, so that they are unsuitable for converting large scale datasets or processing real-time signals.

Motor imagery (MI) is a valuable paradigm of brain-computer interface (BCI). It is a process to imagine activities without actually performing movement. Previous studies have proved that MI shares the same neural circuits with actual exercise process [5]. The classification task of motor imagery EEG signals is important to assist the development of BCI equipment for people with disabilities, promote the rehabilitation of stroke patients. It also has broad application values in preventing sports fatigue. However, because the commonly used EEG signal has low signal-to-noise ratio and obvious temporal and spatial characteristics, the past ANN or SNN method often cannot directly extract the relevant features in EEG signals [6]. Therefore, it is often necessary to develop new methods for extracting the characteristics of data with temporal information [7–10]. However, although that SNN naturally has the ability to process temporal information [9, 11–14], the existing works pay less attention to the combination of SNN and BCI tasks, and always limited to some simple scenario such as seizure detection [7, 9, 11], sleeping [15] and two classification of motor imagery [4, 8].

Classifying an EEG-based four classes motor imagery task still remains a challenging task. In addition, due to inter-subject differences, the performance of cross-subject classify-cation is also poor., It is promising to apply SNN in MI classification tasks, and may further expand the role of SNN in BCI, given that SNN has the advantages of low energy consumption and efficiency in processing of temporal information.

In this study, a parameter-wise gradient descent (PW-GD) method is firstly proposed to optimize the relevant parameters of BSA, this method greatly improves the optimization efficiency under the premise of maintaining the conversion effect. Secondly, we use LSTM to preliminarily filter the spike trains, and then combine a four layers SNN to extract temporal information and carry out the four classification tasks. Meanwhile, we apply surrogate gradient function and channel-wise normalization in training process to ensure the effectiveness of network training. The proposed network architecture was verified on the Physionet dataset by conducting baseline training on 20 subjects to get the pre-trained model and transferring the pre-trained model to a single subject. The effect of cross-subject transfer training on 30 subjects achieved an average accuracy of 65.75%. As far as we know, this is the first time that directly-trained SNN achieves the similar results with CNN in four classes MI task.

2 Methods

2.1 BSA Based on Parameter-Wise Gradient Descent Optimization Method

The core idea of BSA algorithm is comparing the error terms caused by two possible states, i.e., firing spike or not, and then judge whether a spike should be fired at current time [3]. The two error terms are as follows:

$$\begin{cases} err_1 = \sum_{k=0}^{N} |s(k+\tau) - h(k)| \\ err_2 = \sum_{k=0}^{N} |s(k+\tau)| \end{cases} \tag{1}$$

In the above formulas, N represents the length of finite impulse response filter (FIR), $h(k)$ represents spike response of FIR in time k, and s represents the estimated stimulation of neurons. The calculation of s can be concluded as follows:

$$s(t) = (h*x)(t) = \sum_{k=1}^{N} h(t - t_k) \tag{2}$$

where $x(t)$ denotes the spike trains of a certain neuron, given by $x(t) = \sum_{k=1}^{N} \delta(t - t_k)$, with $\{t_k\}$ denoting the set of firing times of the neuron.

In the classical BSA method, a spike will be fired when the following formula is satisfied.

$$err_1 < err_2 - threshold \tag{3}$$

To sum up, there are three main parameters that influence the conversion result of BSA, namely cut-off frequency f_{cut} of FIR, the length L of FIR and the threshold θ to determine whether the spike is fire or not. In this work, a parameter-wise gradient descent (PW-GD) method is proposed to optimize the above three parameters, as shown in Algorithm 1. In the PW-GD algorithm, we use criterion function value *cir* for parameter-wise optimization, and finally get the optimal criterion function value cir^* and three optimized parameters. Moreover, in order to further improve the efficiency when maintaining the conversion effect, we set an expected value cir^t to end the optimization in advance. The algorithm does not pursue the best, and can avoid some unnecessary operation iterations to further accelerate the optimization speed.

2.2 LIF Model and ALIF Model

Various neuron models have proposed to construct SNN, such as H-H model, Spiking response model, Izhikevich model, LIF model and so on. Among all these models, the LIF model is widely used in the construction of SNN as it takes both biological reliability and computational efficiency into account. The membrane update formulas of LIF model are as follows:

$$u_i^t = \left(u_i^{t-1}\alpha + \sum_j w^{ij} s_j^t\right)\left(1 - s_i^t\right) + u_r s_i^t \tag{4}$$

$$s_i^t = f_s\left(u_i^t, \theta_i\right) = \begin{cases} 1, \ if \ u_i^t > \theta_i \\ 0, \ otherwise \end{cases} \tag{5}$$

In above formulas, the u_i^t represents the membrane potential of neuron i at time t, and α equals to $\exp(-\Delta t/\tau_m)$, which indicates the decay factor of membrane potential.

Algorithm 1 PW-GD Optimizing

Input: $EEG, f^0_{cut}, L^0, \theta^0, lr, cir^t$
Parameter: Limits of L, i.e., L_u and L_l
Output: $f_{cut}, L, \theta, cir^*$
1: Let $cir^* = 0$, $f_{cut} \leftarrow f^0_{cut}$, $L \leftarrow L^0$, $\theta \leftarrow \theta^0$.
2: Set $\Delta = 0.001$
3: **While** epoch<100 and $cir^* < cir^t$ **do**
4: $cir^1 \leftarrow BSA\_unity(EEG, f_{cut}, \theta, L)$
5: **if** $cir^1 > cir^*$ **then**
6: $cir^* \leftarrow cir^1$
7: end **if**
8: $f'_{cut} \leftarrow f_{cut} + \Delta$
9: $cir^2 \leftarrow BSA\_unity(EEG, f'_{cut}, \theta, L)$
10: $grad_1 \leftarrow (cir^2 - cir^1)/\Delta$
11: $f_{cut} \leftarrow f_{cut} - lr * grad_1$
12 $\theta' \leftarrow \theta + \Delta$
13: $cir^3 \leftarrow BSA\_unity(EEG, f_{cut}, \theta', L)$
14: $grad_2 \leftarrow (cir^3 - cir^1)/\Delta$
15: $\theta^* \leftarrow \theta - lr * grad_2$
16: change the lr
17: end **While**
18: **for** L in $[L_l, L_u]$ **do**
19: $cir^4 \leftarrow BSA\_unity(EEG, f^*_{cut}, \theta^*, L)$
20: **if** $cir^4 > cir^*$ **do**
21: $cir^* \leftarrow cir^4$
22: end **if**
23: end **for**
24: **function** $cir = BSA\_unity(EEG, f_{cut}, \theta, L)$
25: $Spike, fir \leftarrow BSA(EEG, f_{cut}, \theta, L)$
26: $cir \leftarrow BSA\_Decode(EEG, Spike, fir)$
27: **return** cir
28: **end function**

The Δt is a fixed sample time and the decay time constant τ_m is a learnable parameter. The connection weight between presynaptic neuron j and postsynaptic neuron i can be expressed as $w^{i,j}$. The rest voltage of neurons is represented by u_r. And s^t_i represent the spike emitted by neuron i, the θ_i denotes the spike firing threshold of neuron i, and this value always is diverse for each neuron.

In order to make a single neuron have higher expressiveness, so as to improve the overall performance of SNN, an adaptive LIF (ALIF) model had been proposed. This neuron model has higher feature extraction ability by deploying such neurons in the back layer network of SNN [12, 16]. The membrane update formulas of ALIF are the same as those of LIF. The difference between ALIF and LIF is that the spike firing threshold θ_i will be dynamically adapted at the time of a spike fired. The adapting equations of θ_i are as follows

$$\eta^t_i = \eta^t_i e^{-\frac{\Delta t}{\tau_a}} + \left(1 - e^{-\frac{\Delta t}{\tau_a}}\right) s^{t-1}_i \tag{6}$$

$$\theta^t_i = \theta^0_i + 1.8\eta^t_i \tag{7}$$

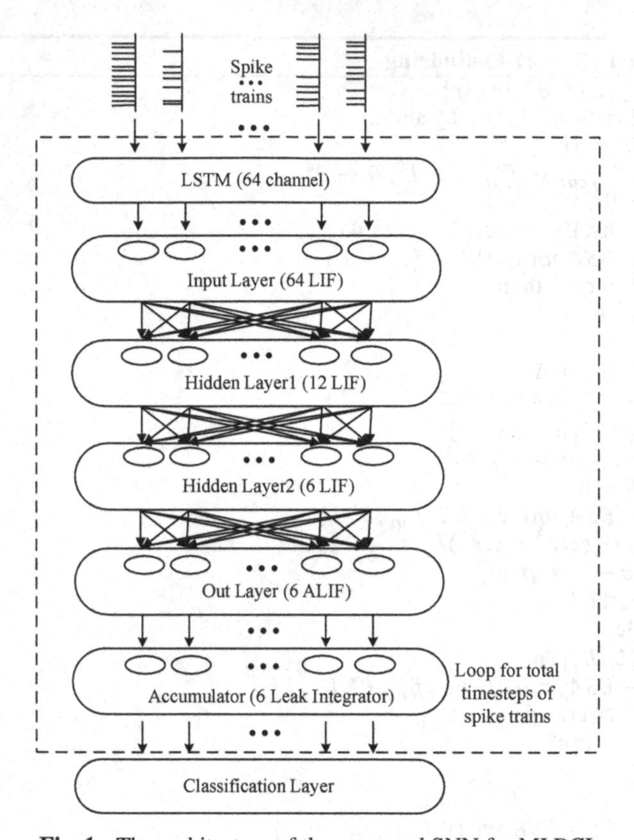

Fig. 1. The architecture of the proposed SNN for MI BCI.

where η_i^t represents the adapting regulator of neuron i at time t, this parameter is related to the spiking trains fired by neuron i. Meanwhile, the regulator itself will decay over time, and the decay constant τ_a is a trainable parameter, which determines the speed of threshold attenuation.

By increasing the release threshold after each spike firing, ALIF can reduce the firing of multiple spikes in a short time and generate more sparse spike trains, thus transmit more accurate temporal information and be expected to improve the overall robustness and performance of SNN.

2.3 The Architecture of SNN

As presented in Fig. 1, the proposed structure consists of one layer of spiking LSTM and four layers of SNN, where the SNN layers are linked through full connections (Fc), and the other connections are paired connections. After receiving and processing all spike trains, the network outputs the data of the accumulator to the classification layer.

2.4 Surrogate Gradient

Due to the nonlinearity and non-derivation of the spike firing formulas of neurons in SNN, gradient backpropagation algorithm cannot be directly used in SNN training. In order to achieve effective and efficient training of SNN, lots of supervised learning methods such as SpikeProp [17], Rrop [18], ReSuMe [19] and Multi-Resume [20] have been proposed. However, due to the limitation of these algorithms on the number of neurons' spike firing, it is difficult to bring out large-scale and continuous training.

The breakthrough appears in the thought that replace the derivative of spike firing function in spiking neurons by derivative of a continuous function [17]. In this way the backpropagation method can be applied to the training process of SNN. This method is called surrogate gradient method [21]. It has been proved that an effective supervised learning method can be used to construct high-performance SNN [14, 15, 21, 22]. The commonly used surrogate gradient functions are shown in Fig. 2.

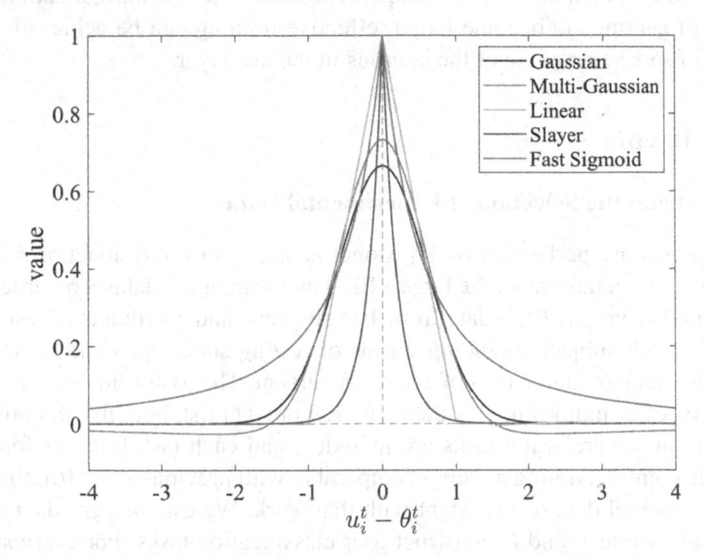

Fig. 2. Schematic diagrams of commonly used surrogate gradient functions.

2.5 Channel-Wise Normalization

In SNN, the neurons accumulate membrane potential by receiving spike trains, and the postsynaptic neurons require more spikes from post layer to surpass the threshold and fire a spike. Therefore, with the deepening of the SNN, the neurons in deeper layer will fire fewer spike. This phenomenon is called spike degradation and may result in invalid training. Considering that we use a four-layer SNN, it is necessary to apply effective methods to avoid spike degradation.

Channel-wise normalization has been proved to be an effective method for constructing deep SNN [14]. Compared to layer-wise, channel-wise normalization focuses on the difference of spike-firing rate between each pair of connected neurons rather than the whole layer, thus it is much more flexible and accurate. The formula for normalization used in this method is as follows. The spike-firing rates of presynaptic neuron j and postsynaptic neuron i are denoted by λ^j, λ^i respectively, the parameters are decimals between 0 and 1.

$$w^{i,j} = w^{i,j} \times \frac{\lambda^i}{\lambda^j} \tag{8}$$

$$b^i = \frac{b^j}{\lambda^j} \tag{9}$$

Obviously, the channel-wise normalization cannot be carried out in every epoch, otherwise the value of weights and biases (especially the latter) will continue to increase, leading to invalid training. Thus, in the training process, we normalize the value of weights and biases only once after five epochs of training. The purpose of delaying five epochs is to make the model slightly adapt to the data. After the normalization, although the values of parameters become larger, effective training can be achieved due to the increasing of spike-firing rate of the neurons in the last layer.

3 Experiments

3.1 Dataset and the Selection of Experimental Data

The experiments are performed on Physionet dataset (www.physionet.org/content/eegmmidb/1.0.0/), which is one of the largest EEG motor imagery dataset recorded by BCI-2000 system. It contains EEG data from 109 subjects, and 14 rounds of test data were collected for each subject, including 2 runs of resting state (open and closed eyes), 6 runs of motor imagery and 6 runs of actual movement. The motor imagery tasks include 4 different types of motor imagery data, i.e., left or right fist, both fist and both feet.

In each run, several valid tasks are included, and each task lasts for four seconds (640 sample points). Aiming at being comparable with previous work [6], the selection of our experimental data is consistent with that work. We use imagery data of left fist, right fist and both feet, and to construct four classification tasks. For every subject, 84 tasks (21 tasks in each class) are selected, and for each task, we use the data of 3 s (480 sample points) after the task cues appear.

3.2 The Performance of PW-GD Optimizing for BSA

In order to test the efficiency and effectiveness of the proposed PW-GD optimization algorithm. We select 60 channels of EEG data from 5 subjects as the original samples, and the BSA optimized by the three optimization algorithms, i.e., grid search (GS), generation algorithm (GA) and PWGD, are used to convert those data respectively.

We use signal-to-noise ratio (SNR) as criterion for optimization and comparison between algorithms. The calculation formula of SNR can be concluded as follows:

$$SNR = 20\ log_{10} \frac{\sum_{k=1}^{N} |S_O(k)|^2}{\sum_{k=1}^{N} |S_O(k) - S_c(k)|^2} \tag{10}$$

Table 1. Comparison of the average performance of the three optimization algorithms (per channel), n_{BU} indicates the number of executions of the function *BSA_Unity*.

Name	epoch	n_{BU}/ epoch	Run time (Sec)	Run time (unitized)	SNR (dB)
GS	1	$50 \times 50 + 20$	23.94	45.17	37.92
GA	30 (fixed)	$2 \times 60 + 20$	54.86	103.51	35.26
PWGD (Ours)	1	42.2 (avg.) + 20	**0.53**	1	**38.25**

Table 2. The architecture of LSTM-SNN, N_b denotes the batch number.

Name	Layers	Components	Params number	Output shape
SNN	LSTM	$4(w_{ih}, b_{ih}, w_{hh}, b_{hh})$	33280	$N_b \times 64 \times 1$
	Input	64 LIF (τ_m, θ)	64×2	
	Fc 64–12	w, b	$12 \times 64 + 64$	$N_b \times 12 \times 1$
	Hidden 1	12 LIF (τ_m, θ)	12×2	
	Fc 12–6	w, b	$6 \times 12 + 12$	$N_b \times 6 \times 1$
	Hidden 2	6 LIF (τ_m, θ)	6×2	
	Fc 6–6	w, b	$6 \times 6 + 6$	
	Output	6 ALIF (τ_m, τ_a, θ)	6×3	
Classifier (execute only once)	Accumulator	6 LI $(\tau_m^{'})$	6	$N_b \times 6$
	BN	\	0	$N_b \times 6$
	Fc 6–4	w, b	$4 \times 6 + 6$	$N_b \times 4$
	Decision	Log-softmax + max	0	N_b
Total			34456	

In the above formula, $S_o(k)$ and $S_c(k)$ represent the value of the original signal and the converted signal at time k respectively, and N represents the length of EEG signal.

Table 1 shows the mean conversion performance of BSA under the three optimization algorithms, in term of the run time and SNR. In order to keep the comparability of the three algorithms, the main parameters to be optimized in the three algorithms are f_{cut} and θ. As for the integer L, we uniformly search the best value in the range of 10–30. From the results shown in Table 1, we can clearly see the superiority of PW-GD algorithm. That is, compared with the previous two algorithms, the efficiency is improved by nearly 50 and 100 times respectively when retaining the quality.

Meanwhile, it is necessary to discard some imperfect data to ensure the validity of training. The criteria adopted to select data consist of the integrity of events and the and the whole last time of data from every subject. In this way, we discard three incomplete subjects (S88, S92, S100, S104) (Fig. 3).

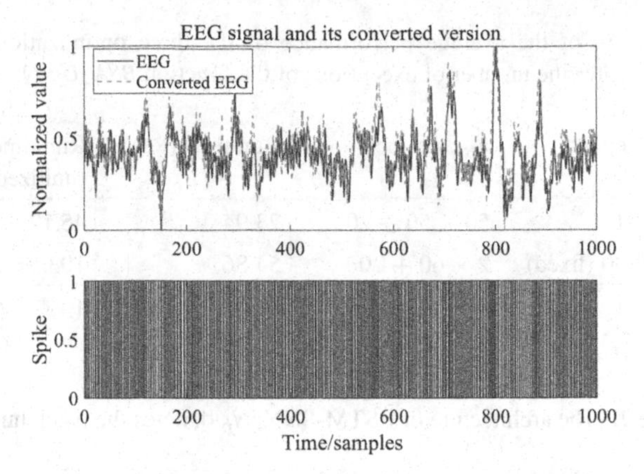

Fig. 3. The EEG signals were encoded into the spike trains. Top: the contrast of EEG and converted EEG (SNR = 37.2) Blow: the spike train obtained with the optimal BSA.

Table 3. Model hyperparameters used in experiment

Parameter	Value
Optimizer	Adam
Learning rate	4e-4
τ_m	20
τ_a	7
θ_0	0.06
L1 regularization	1e-5
Ratio of dropout	0.5
Batch size	64
Surrogate gradient	Slayer
Training epochs	150 (baseline)/10 (transfer)

3.3 The Performance of SNN in MI Classification

The proposed SNN structure is summarized in Table 2, while the hyperparameters of training process are provided in Table 3. The training process is consisted of two parts: the baseline training and the transfer training on a single subject.

We first used the data from 20 subjects for training baseline model, these data are randomly divided into a training set and a test set in the proportion of 80/20. After 150 epochs of training, the baseline model will be used as the pretrained model. The curves of the accuracy and the values of loss function during 150 epochs are shown in Fig. 4. It can be found that no overfitting phenomenon appear by introducing L1 regularization and dropout.

Next, we transfer the pretrained model on the data of one single subject. Note that the subject is not included in the 20 subjects used in baseline training. For transfer training, the data of the single subject are randomly divided into 75/25, consistent with the previous CNN method. The largest difference between the transfer training and the baseline training is the number of training epoch. Transfer training allows only ten iterations to adapt to a new subject, aiming to achieve accurate and rapid data recognition based on small samples. After ten epochs of training, the test accuracy of the tenth epoch will be viewed as the recognition accuracy of the current subjects. We conduct transfer training on 30 subjects and the distribution of accuracies is shown in Fig. 5.

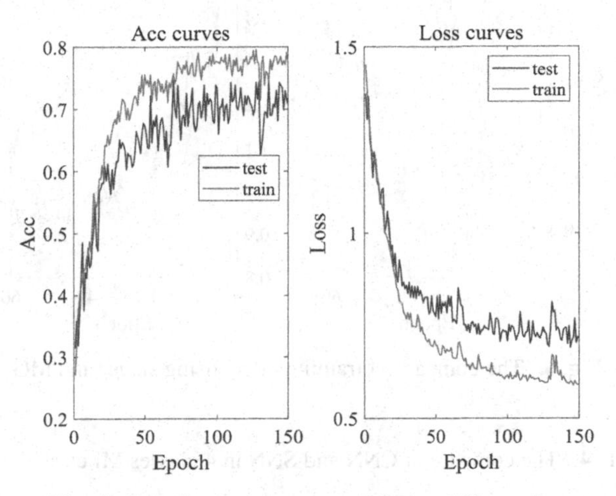

Fig. 4. Baseline training curves during 150 epochs.

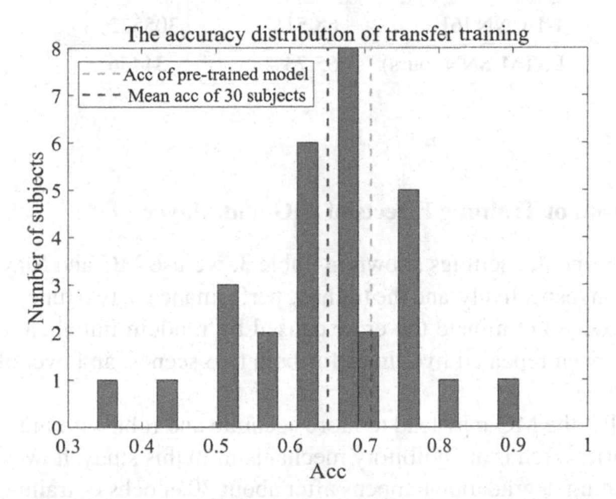

Fig. 5. The distribution of the classification accuracies of

The experimental results show that our baseline model reaches the cross-validation accuracy of 70.8%, and the average accuracy of transfer training on 30 subjects is 65.8%, as presented in Table 4. This result is comparable to the previous work based on 1D-CNN and fairly good for EEG-based four classes MI classification task, and the scale of the model is much smaller. By further optimizing the network structure and adopting more effective training methods, the effect of SNN can be further improved.

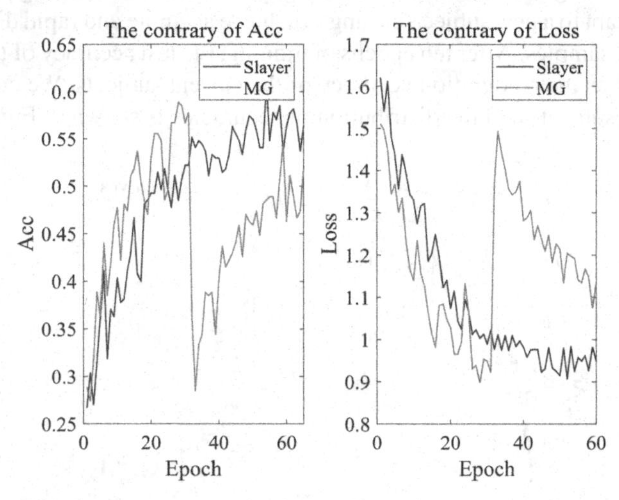

Fig. 6. The compare of training effect using slayer and MG.

Table 4. The compare of CNN and SNN in 4 classes MI classification.

Name	Subject acc.	Parameters
1-DCNN [6]	**68.51%**	305522
LSTM-SNN (ours)	65.75%	**34456**

3.4 Comparison of Training Effects of MG and Slayer

Under the experimental settings shown in Table 3, we use MG and Slayer as surrogate gradient function respectively, and the training performance of two functions is compared in Fig. 6. In order to eliminate the error caused by random initialization, the training processes have been repeated five times for both two scenes, and overall trends remain basically the same.

Theoretically, the MG may lead to more accurate and robust training by introducing the negative part, which is an inhibitory mechanism. In this study, however, we find that an obvious training degradation happens after about 30 epochs of training in MG scene. This phenomenon will still occur after changing the height or width of MG function.

Even if we reduce the learning rate, the phenomenon has only been alleviated. On the other hand, in the Slayer scene, there is basically no obvious degradation in the network training. Therefore, we speculate that the degradation is caused by the negative gradient part of MG, and MG function may not suitable when processing EEG data. Definitely, more experiments and analysis are required to validate this speculation.

4 Conclusion

In this study, we first proposed PW-GD optimizing method for BSA. This method greatly improves the conversion efficiency of spiking encoding in EEG signals whereas maintaining great quality. Consider that there are notable differences between subjects in EEG signals, it is necessary to optimize the parameters of BSA for each subject. This method can used to efficiently convert large-scale EEG datasets into spike shape datasets and then promote the application of SNN in BCI. On the other hand, it can also significantly accelerate the real-time processing of EEG signals by SNN and thus expand the online application of SNN.

Secondly, a LSTM-SNN architecture is developed for 4-classes MI classification task. Based on this SNN structure, the mean accuracy of 65.8% has been achieved across 30 subjects, it is comparable with previous work based on the 1D-CNN, and the amount of network parameters is only 1/10 of 1D-CNN.

Furthermore, the proposed framework can also be applied to other EEG-based classification tasks such as emotion recognition, sleep recognition, autistic spectrum disorder (ASD) diagnosis, as the temporal information of these tasks can be reserved by BSA and be extracted by LSTM-SNN.

Although our work demonstrates the potential of SNN in complex EEG classification tasks, the effect of this work is far from practical application and still need to be improved. Our next work will focus on the optimization of network structure, such as introducing neurons with more dynamic characteristics, and proposing more effective normalization method or network connection mode to obtain deeper SNN.

References

1. Davies, M., et al.: Loihi: a neuromorphic manycore processor with on-chip learning. IEEE Micro **38**(1), 82–99 (2018)
2. Hu, Y., Li, G., Wu, Y., Deng, L.: Spiking neural networks: A survey on recent advances and new directions. Control and Decision **36**(1), 1–26 (2021)
3. Schrauwen, B., Campenhout, J.: BSA, a fast and accurate spike train encoding scheme. In: Proceedings of the International Joint Conference on Neural Networks, vol. 4, pp. 2825–2830 (2003)
4. Nuntalid, N., Dhoble, K., Kasabov, N.: EEG classification with BSA spike encoding algorithm and evolving probabilistic spiking neural network. In: Lu, B.-L., Zhang, L., Kwok, J. (eds.) ICONIP 2011. LNCS, vol. 7062, pp. 451–460. Springer, Heidelberg (2011). https://doi.org/10.1007/978-3-642-24955-6_54
5. Lotze, M., Halsband, H.: Motor imagery. J. Physiology-Paris **99**, 386–395 (2006)
6. Dose, H., Moller, J.S., Iversen, H.K., Puthusserypady, S.: An end-to-end deep learning approach to MI-EEG signal classification for BCIs. Expert Syst. Appl. **114**, 532–542 (2018)
7. Ghosh-Dastidar, S., Adeli, H.: A new supervised learning algorithm for multiple spiking neural networks with application in epilepsy and seizure detection. Neural Network **22**(10), 1419–1431 (2009)

8. Carlos, D., Juan, H., Antelis, J.M., Falcon, L.E.: Spiking neural networks applied to the classification of motor tasks in EEG signals. Neural Netw. **122**, 130–143 (2020)
9. Wang, Q., Wang, L., Xu, S.: Research and application of spiking neural network model based on LSTM structure. Appl. Res. Comput. **38**(5), 1381–1386 (2021)
10. Wang, Z., Zhang, Y., Shi, H., Cao, L., Yan, C., Xu, G.: Recurrent spiking neural network with dynamic presynaptic currents based on backpropagation. Int. J. Intell. Syst. **37**(3), 2242–2265 (2021)
11. Buteneers, P., Schrauwen, B., Verstraeten, D., Stroobandt, D.: Real-time epileptic seizure detection on intra-cranial rat data using reservoir computing. In: Köppen, M., Kasabov, N., Coghill, G. (eds.) ICONIP 2008. LNCS, vol. 5506, pp. 56–63. Springer, Heidelberg (2009). https://doi.org/10.1007/978-3-642-02490-0_7
12. Bellec, G., Salaj, D., Subramoney, A., Legenstein, R., Maass, W.: Long short-term memory and learning-to-learn in networks of spiking neurons. In: Advances in Neural Information Processing Systems, pp. 787–797 (2018)
13. Kim, Y., Panda, P.: Optimizing deeper spiking neural networks for dynamic vision sensing. Neural Netw. **144**, 686–698 (2021)
14. Kim, S., Park, S., Na, B., Yoon, S.: Spiking-YOLO: spiking neural network for energy-efficient object detection. Proc. AAAI Conf. Artif. Intell. **34**(7), 11270–11277 (2020)
15. Jia, Z., Ji, J., Zhou, X., Zhou, Y.: Hybrid spiking neural network for sleep electroencephalogram signals. Sci. China Inf. Sci. **65**, 140403 (2022). https://doi.org/10.1007/s11432-021-3380-1
16. Yin, B., Corradi, F., Bohté, M.: Accurate and efficient time-domain classification with adaptive spiking recurrent neural networks. Nature Machine Intelligence **3**(10), 905–913 (2021)
17. Bohte, S.M.: Error-backpropagation in networks of fractionally predictive spiking neurons. In: Honkela, T., Duch, W., Girolami, M., Kaski, S. (eds.) ICANN 2011. LNCS, vol. 6791, pp. 60–68. Springer, Heidelberg (2011). https://doi.org/10.1007/978-3-642-21735-7_8
18. Mckennoch, S., Liu, D., Bushnell, L.G.: Fast modifications of the spikeprop algorithm. In: Proceedings of the 2006 IEEE International Joint Conference on Neural Network, pp. 3970–3977 (2006)
19. Ponulak, F., Kasinski, A.: Supervised learning in spiking neural networks with ReSuMe: sequence learning, classification, and spike shifting. Neural Comput. **22**(2), 467–510 (2010)
20. Taherkhani, A., Belatreche, A., Li, Y., Maguire, L.: Multi-DL-ReSuMe: multiple neurons delay learning remote supervised method. In: 2015 International Joint Conference on Neural Networks, pp. 1–7 (2015)
21. Neftci, E.O., Mostafa, H., Zenke, F.: Surrogate gradient learning in spiking neural networks: bringing the power of gradient-based optimization to spiking neural networks. IEEE Signal Process. Mag. **36**(6), 51–63 (2018)
22. Han, B., Srinivasan, G., Roy, K.: RMP-SNN: residual membrane potential neuron for enabling deeper high-accuracy and low-latency spiking neural network. In: Proceedings of the IEEE/CVF Conference on Computer Vision and Pattern Recognition, pp. 13558–13567 (2020)

Virtual Drone Control Using Brain-Computer Interface Based on Motor Imagery Brain Magnetic Fields

Gaobo Tan, Jinming Gai, Ruihan Guo, Guiying Zhang, Qiang Lin, and Zhenghui Hu(✉)

College of Science, Zhejiang University of Tecchnology, Hangzhou, China
lxy@zjut.edu.cn
http://www.lxy.zjut.edu.cn/index.php

Abstract. Brain-computer interfaces (BCIs) based on brain magnetic fields is a novel trend in the field of rehabilitation robotic that could be leveraged for helping the patients who have lost voluntary muscle control to communicate. This study present an experiment of controlling a virtual drone in three-dimensional space using brain magnetic fields induced by left or right hand motor imagery. We applied optically-pumped magnetometers (OPMs) to capture brain magnetic fields, Subjects sat in the magnetically shield room (MSR) comfortably and performed the motor imagery (MI) task corresponded to the cue provided via an brain magnetic fields based BCI system whilst saving the data attached with variable classlabels. Then, we processed the data in time and frequency domain in order to extract the signal feature in form of an event-related desynchronization (ERD) or event-related synchronization (ERS). Furthermore, machine learning was employed to be the identification tool for motor imagery brain magnetic fields and further, controlled a virtual drone flight according to commands.

Keywords: Brain-computer interface · Motor imagery · Brain magnetic fields · Machine learning · Control virtual drone

1 Introduction

Brain magnetic fields based Brain-computer interfaces is a novel conception which is endowed with promising potential in the field of medical rehabilitation that enable a man with paralysis to communicate through brain activity by using a non-invasive manner. several kinds of events, such as open-close eyes induction, auditory evoked test and motor imagery, can result in inducing time-locked changes in the activity of neuronal populations that are generally called event-related potentials (ERPs). Indeed, it has been proved that the evoked activity, or signal of interest, has a more or less fixed time-delay to the stimulus, and the evoked potentials (EPs) can be considered to result from a reorganization of the phases of the ongoing brain magnetic fields [1]. This mean that

X. Ying (Ed.): HBAI 2022, CCIS 1692, pp. 161–171, 2023.
https://doi.org/10.1007/978-981-19-8222-4_14

the amplitude of these ERPs in specific frequency band represent changes of the ongoing brain activities, in narrow sense, either of decreases or of increases of power in given frequency bands, the former term is named ERD while the latter term named ERS.

In general terms, brain magnetic fields has traditionally been detected by the superconducting quantum interference devices (SQUID). However, there are an amount of drawbacks of SOUID such as immobile and expensive. Conversely, resent studies demonstrate that OPMs have the potential to overpass many of the limitations of SOUID, for instance, OPMs could work in the room temperature and be placed closer to the scalp than SOUID, enable to detect subtler signal with high spatial resolution [2,3]. Furthermore, since the flexibility of placement means OPMs array can, in principle, be adapted to whole head. Consequently, due to the reason of, sometimes, sensors get too close to the scalp, head oscillation would result in generating motion artifact distort when measure brain activities. Therefore, the development of lightweight ergonomic helmets, was able to adapt in diverse people, could alleviate the impact of motor oscillation, has significant sense in experiment [4].

In this work, in order to achieve our experiment purposes, we chose to use commercially available zero-filed OPM sensors QuSpin, sensors were placed closely to the scalp and fixed at purpose-made holder on the rigid additively manufactured helmet we designed. The helmet was mounted at a wooden platform which was cling to MSR wall. Three sensors were placed on C3, C4 and Cz positions individually corresponded to the 10–10 International Standard, that they nearly cover the area of sensorimotor cortex of left and right hand movement. Subjects would sat in a comfortable chair in MSR and wear the helmet, then, executed two different task, performed right or left-hand motor imagery (repeat grasp closing-relaxing). The commands were displayed via luminous diodes hung on the wall both side. After that, we processed the data our collected and explore its performance on various parameters offline. Some suitable data would be picked out and grouped by mean of machine learning, and then, encode the control signal applied in controlling virtual drone.

2 Brain Magentic Fields Based BCI System Description

The preparation phase of this study consist of experimental environment construction and subject training. Subjects with no known history of neurological or psychiatric disorders including claustrophobia (subject need to stay in MSR about twenty minutes). Each subject provided written consent in order to participate in a protocol. In the following subsections, we demonstrate the our work in helmet design (Sect. 2.1) and experimental environment construction (Sect. 2.2).

2.1 Helmet Design

It is a critical point to the ultimate design of a whole MI based BCI system that is the method how OPM sensor are mounted on the head. However, there are no available commercial ergonomic helmet can be adapted to this study, we decide to make it depend on study requirements. According to the research demands, this design ought to represent a balance of five critical considerations: first, helmet must cover most of brain activity regions relative to brain electrical activity mapping (BEAM) by right of the existing brain anatomy study. Second, in order to make sure that sensors pick up maximum signal and avoid artefact in measurement, the sensors should to be placed close to the head as possible and rigidly held in position and no movement relative to head. Third, due to the sensor's vapor cell is electrically heated with resistive heaters to around 150 °C and need to be dynamically stabilized at 10mK level, it is vital to considerate the heat-sinking capability of helmet in design. Further, it should be taken account of the ergonomic performance of helmet to ensure subject feel comfortable, having significant sense in removing motion artifact. Last but not least, the helmet aim to be adapted for the majority of normal size people and, the best is, to be designed into embedded style that we could take it down and move it to other place when we need.

For our purposes, we choose to employ 3D printing technology in making rigid helmet. The helmet is manufactured from a kind of nylon polymer named PA12, white, with no unpleasant odor, and harmless to human. The size and shape of the inner surface is based on metrical data measured from subjects who already join in. The outer surface of helmet is 7mm apart from the internal surface, that will make helmet not to be out of shape easily. The area on the helmet correspondent to ear has been cut off in order to improve its comfort and wearability. The helmet contains 78, $15.6 \times 19.8\ mm^2$ square sockets, visibly, those sockets are divided into three separately arrays, one of them is placed at the region in which parietal lobe and temporal lobe relative to whilst one is sited at another side, the rest one is placed at occipital lobe. However, despite those methods we use to make helmet close to scalp surface as possible, there are still tiny gap existing among them. To solve this problem, in addition, we design a sort of small mounts manufactured by plain, white resin, which could clamp OPM sensor and be contained in square socket, enabling shorten the gap between sensors and scalp by pushing sensor inward. There is the picture "Rigid Additively Manufactured Helmet with socketst", show as Fig. 1(a).

2.2 Experimental Environment Construction

Magnetoencephalography (MEG) studies show that, 10 fT-1 pT magnetic fields produced by electric currents flowing in neurons [5]. Due to the superior difficulty of capturing such tiny signal, the critical joint to bridge brain motor-imagery activity with machine controlled operation is that, fabricate an adaptable, motion-robust brain magnetic measured environment. This section expand upon the factors of experimental environment construction from aspects of instruments installation in MSR and host program modification.

Instruments. Under existing conditions, we fullfill the brain magnetic signal measured task by using three zero-field, second generation magnetometers manufactured by QuSpin named QZFM sensor. The typical magnetic field sensitivity

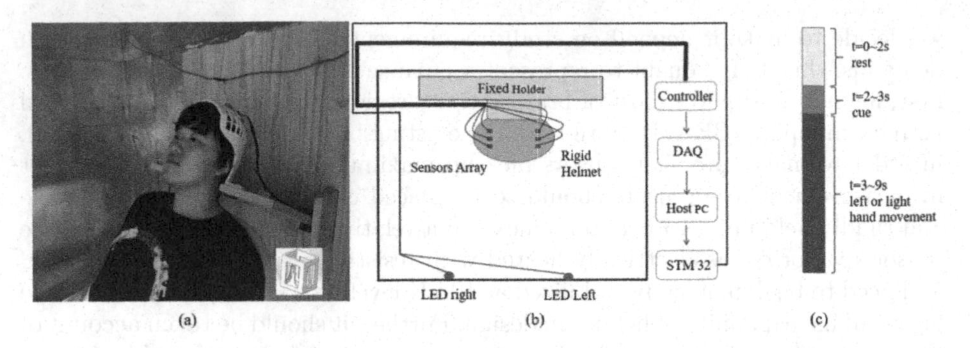

Fig. 1. The experiment set: (a) Rigid additively manufactured helmet with sockets; (b) General view of MI based system; (c) Experimental paradigm.

of QZFM is 7–13 $\mathrm{fT}/\sqrt{Hz}$ in 1–100 Hz frequency band [6], In addition, the sensor response is linear to within 1% when the magnetic signal amplitude is less than 1 nt, satisfy the need of brain magnetic fields measurement. The outer shape parameter of the sensor is $12.4 \times 16.6 \times 24.4$ mm, the 6.5 m signal transmission cable connecting the sensor and the electronics controllor[1], allow electronics controllor placed away from its sensor unit.

Because zero field OPMs operate correctly only under ground of low intensity megnetic filed, therefore, we applied sensors in a $1.6 \times 2.2 \times 1.9\,\mathrm{m}^3$ MSR to competely cut off outside interference[2]. The MSR comprise multiple permalloy-layer, north and south-oriented wall install three-circle degaussing coils, and that, this rear wall have a hole 10 cm in diameter used for communicate cable to across the wall. A wooden holder cling to the MSR wall, thus, helmet could fixed on it, minimizing the adverse affect of random oscillation outside and subject's unconscious movement. We rest a plastic chair in front of wooden holder, which can make subject feel comfortable during experimental session.

For the purposes of this study, left-right directional command signals were sent to subject randomly along with a constant time interval, therefore, the subjects were able to complete the motor-imagery task in due time. On the left hand side and right hand side, we set up a pair of LED diode on the wall at a proper height. Using a single chip microcomputer (SCM) STM32 to lighten either diode and dim another one as cue, subjects are supposed to perform motor-imagery task correspondently at a same time. STM32 connect with host computer approach by universal synchronous asynchronous receiver transmitter (USART) follow the transmit protocol we design, when SCM receive the string "LEFTLEDON" from host computer, then lighten the left side LED. In contrast, if SCM receive "RIGHTLEDON", lighten the right side LED. Else, none of them lighten.

[1] The bridge between QZFM sensor and host computer is UART to USB serial port line.

[2] The height of MSR is 1.6 m.

Host Control Program. For controlling the magnetometer, we installed the control software QZFM version 0.3.1 provided by QuSpin office on host computer[3]. The software would benefit host computer to generate magnetic modulation signal used for lock-in detection so that ensure the sensors work in zero-field. QZFM electronics controller could digitized the analog output by a digital signal processor (DSP) automatically. Thus, it is very convenient that the software are able to show magnetic signal directly. Furthermore, we applied Data Acquisition (DAQ) device (NI-9239, manufactured by National Instruments) in changing analog output from electronics controller into digital data form, and then transmitted to host computer. Next step, we used LabVIEW for programming to read and save the signal data from DAQ. Moreover, this program allow to simple analyse brain magnetic fields data in time-frequency domain and display the information contained in it. So that, when we measured brain magentic signal, it was possible to know whether sensors have got the correct signal we want meanwhile. We made the interactive animations of the drone using Unity 3D software in order to demonstrate the movement of drone based machine learning on host computer offline model, and we will specified the procedure in following machine learning section.

As mentioned above, we construct an integrated MI based BCI system, there is the picture represent the whole procedure, Fig. 1(b).

3 Experiment Content and Data Processing

The data collection and processing task consisted of subject training phase, experiment phase and data analysis phase. More information of this three phases are written in the section.

3.1 Subject Training

The initial target of training was to familiarize the subject with the experimental paradigm and improve the signal-to-noise ratio (SNR). Six able-bodied participants took part in the experiment (both of them are male). Subjects were sat on a chair in MSR without any magnetic material carried, and wear the rigid helmet. And then, they were introduced to and trained in left or right hand movement task, mitigated the motion extent gradually till artifact do not be represented in QZFM. Thus, subjects were shown two visual cues: "clench the fist" and "relax the fist", displayed in form of LED lighting-up and lighting-off. Subject were instructed to repeatedly open and clench left or right hand in a run with 100 trials of 9 s (11 s) length. Visibly, the cue would show at 3 s to 9 s (4 s to 11 s), in contrast, being quiet in remaining time. It is prevalence of performing run 5–10 times for each subject to enhance the SNR.

[3] https://github.com/weinbe58/QuSpin.

3.2 Data Collection and Processing

Data Acquisition. The data of brain magnetic fields was acquired from two subjects picked out among those six volunteers, using three OPMs sensors which were placed into sockets in helmet, at the position C3, C4 and Cz. The experimental paradigm setup is depicted in Fig. 1(c). For the sake of extracting more information for brain magnetic fields during performing motor-imagery task while occupying less host computer resources, the OPM data were recorded 512 Hz sampling rate. The experiment consist of 3 runs with 140 trials each, all runs were conducted on the same day. Moreover, in order to keep subject in decent condition, we set a 30 around minutes break in between. For the same reason, setting the length of per trial to 9s would outperform than 11 s. The first 2 s was quiet, at t = 2 s, the host computer send out an order to SCM, consequently, left or right LED would be lit as a cue. The cue was displayed for 1s, then at t = 3 s, put the LED out and waited for next command. We employed LabVIEW in generating this command, there is a numerical control to produce a variable v in the $-0.5 \leq v \leq 0.5$ window, if the $v > 0$, lighten the left LED and return classlabel "1", in contrast, $v < 0$, the right LED would be lit and return classlabel "2", else, return "0". At same time, the subject was instructed to perform motor-imagery of left or right hand movement guided by LED without real motion. The command of left and right cue was random and evenly distributed. Finally, the raw data was saved in a txt-fileformat, the data set namely **x_data** contains 3 OPMs channels, and wrote their relative classlabels into variable **y_labels**.

Signal Feature Extraction: EDR/EDS. Data analyses conducted on MATLAB, and this dataset was recorded from a normal subject in one day (25-years-old).

In general, the frequency of brain oscillations is negatively correlated to their fluctuation amplitude. With the development of electroencephalogram (EEG) recording and analysis, it's has been confirmed that the signal amplitude within alpha band (8–12 Hz) and low beta band (13–30 Hz) over contralateral sensorimotor cortex would decrease while preparing and performing movements with one hand, suggest that the activate cell assemblies comprise less neurons. In contrast, it shall present increasing trend over homolateral sensorimotor cortex for more neurons be comprised [7,8]. Undoubtedly, ERD/ERS is be considered to be a meaningful method to analyze brain signal function. Therefore, we employed this method in extracting signal feature from raw data.

Firstly, we classified the dataset with 140 trials into two three-dimensional matrixes consisted of sampling points, channels and trials according to the classlabels. Then, we cheblord-filtered all data between 8 12 Hz and conducted squaring. Note that, trials in which the standard deviation of the signal was substantial different across all trails would be regarded as "bad" trials and were removed. After that, we calculated the averaging over whole trials and then conducted smoothing step. Subjects were in rest state at the time between t = 1 s to 2 s per trial, thus, the amplitude of brain activity oscillation under this condition could be considered as reference voltage, and targeted it as the baseline for each trial. Let the A_n denote the value of the processed signal mentioned above for trial n for OPM sensors in C3 and C4, respectively, and R_n represent the mean value of baseline for A_n. Thus we applied a formula in quantification of ERD/ERS:

$$ERD/ERS = \sum_{n}^{3,4} \frac{A_n - R_n}{R_n} \times 100\% \tag{1}$$

We picked a satisfactory dataset and processed it, plotted the ERD/ERS phenomenon of motor-imagery, the result is showed in Fig. 2. Obviously, there is substantial difference between left and right hand motor-imagery, which was in line with our expectations, that not only confirm the utility for the ERD/ERS method, but the validity for the data we measured.

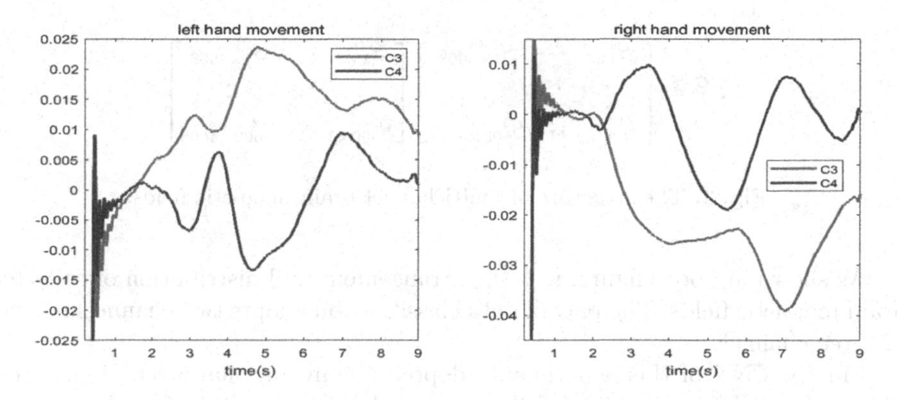

Fig. 2. ERD/ERS phenomenon of left and right hand motor imagery.

4 Machine Learning and Result

Deep learning is one of methods of machine learning that has attracted widespread attention in recent years, and its classical model-Convolutional Neural Network is consider as a very important breakthrough technology. Deep neural network with deep structure can learn features of higher levels of MI (motor imagery) through delamination non-linear mapping [9–11].

It is considered as the most promising tool for extracting the characteristics of brain magnetic fields. Compared with other machine learning models, CNN is particularly suitable for analyzing raw data, CNN can learn the abstract expression of brain magnetic fields. The experiment was not used Two-dimensional convolution, which is better in image processing, because one-dimensional convolution is more effective in signal data processing. Then the trained model is used in practice to distinguish the results of the signal and control the virtual drone.

4.1 Data for Machine Learning

A total of 500 times of motor imagery were used in the experiment. In order to eliminate the interference of other factors as much as possible, the experiment of data was measured in one day. The data of 500 times were used in the experiment for the training set, and a random subject of data of 60 times of motor imagery were employed to used for classification of the testing set by the trained classifier. The brain magnetic fields of motor imagery of left and right hand used in the experiment came from the measurement data mentioned above. The sampling frequency of the signal is 512 HZ, each image is 9 s, and adopts C3 and C4 two-channels. Each generated had generated data of amount is 512×9, and the format of input data is [500, 512×9], it is meaning trains of 500 times with 4608 data in each row. As shown in Fig. 3, there is the processed multichannel magnetic signal, C is the number of channels and T is the coordinate position.

$$S_T^C = \begin{bmatrix} S_{1\times 1}^1 & \cdots & S_{1\times 4608}^1 \\ \vdots & \ddots & \vdots \\ S_{500\times 1}^1 & \cdots & S_{500\times 4608}^1 \end{bmatrix} \begin{bmatrix} S_{1\times 1}^2 & \cdots & S_{1\times 4608}^2 \\ \vdots & \ddots & \vdots \\ S_{500\times 1}^2 & \cdots & S_{500\times 4608}^2 \end{bmatrix}$$

Fig. 3. The structure of multichannel brain magnetic fields.

As shown in above figure, it is the arrangement and distribution of collected brain magnetic fields. The part of data classification adopts two-channels instead of three channels.

Finally, CNN of this experiment adopted 4 convolutional layers, four layers of pooling and four layers of fully connected layer (a total of twelve layers), and there exist pidding in each convolution layer to improve data utilization. BatchNorm1d function (a method to make neural network training faster and more stable) was used. The four pooling layers all adopted maximum pooling layer to prevent overfitting (referring to the phenomenon that the model has high accuracy in the training set but low accuracy in the test set). After convolutional layer and pooling layer, the ReLU activation function was required, the neurons which adopts ReLU only need add, multiply and compare, is more efficient in calculation. ReLU function is also considered to have biological plausibility. We need to add dropout layer after the fully connection layer and specify the drop probability, so that during model training, some nodes of the network will be randomly disabled(the output is set to 0) and the weight will not be updated (but it will be saved and used in the next training). Other processes remain unchanged. Set the dropout layer in the last fully connected layer to prevent overfitting.

4.2 Model Training

After constant adjustment, the parameter structure of the neural network was finally determined. The training process of the network is as follows: CNN loss function is optimized to train the network, and the iteration ordinal number was set to 300. Each row is a train. Because the signal is one-dimensional, the input is also one-dimensional. Adam algorithm is used to train the CNN network, the learning rate is 1×10^{-3}, which can replace the classical stochastic gradient descent, and set a loss rate of 0.5 at the dropout layer, then the whole classification network has been finished. Each row of data input into the training model in turn (Table 1).

Table 1. Details of CNN structure.

Number	Name	Tensor size	Kernel size	Paddig
1	Input	4608×2	–	–
2	Conv	4608×20	9	4
3	MaxPool	1152×20	4	–
4	Conv	1152×30	29	14
5	MaxPool	288×30	4	–
6	Conv	288×50	9	4
7	MaxPool	144×50	2	–
8	Conv	144×30	5	2
9	MaxPool	72×30	2	–
10	Fully Connected	100	–	–
11	Fully Connected	50	–	–
12	Fully Connected	30	–	–
13	Fully Connected	2	–	–

4.3 Test Result

The collected data are more susceptible to noise pollution, but the purpose of this experiment is to classify the unprocessed irregular data sets to increase practicality. So instead of using regular data set data for training, we use our own measured data. The number of iteration is set to 300. The trained model is used to a testing data which come from a random subject to test its accuracy.

When the convergence was finally achieved, the accuracy was 73.3%, which roughly meets the expectation, the result could been seen in Fig. 4. The classifier which was trained link to BCI, and identified real-time data to control virtual drone.

4.4 Virtual Drone Controlling

There is the picture for virtual drone controlling in three dimensional space, as shown in Fig. 5. We ultimately employed machine learning method in changing the condition of virtual drone and making it turn left or right based on the command transmitted from host computer while the relative signal was identified.

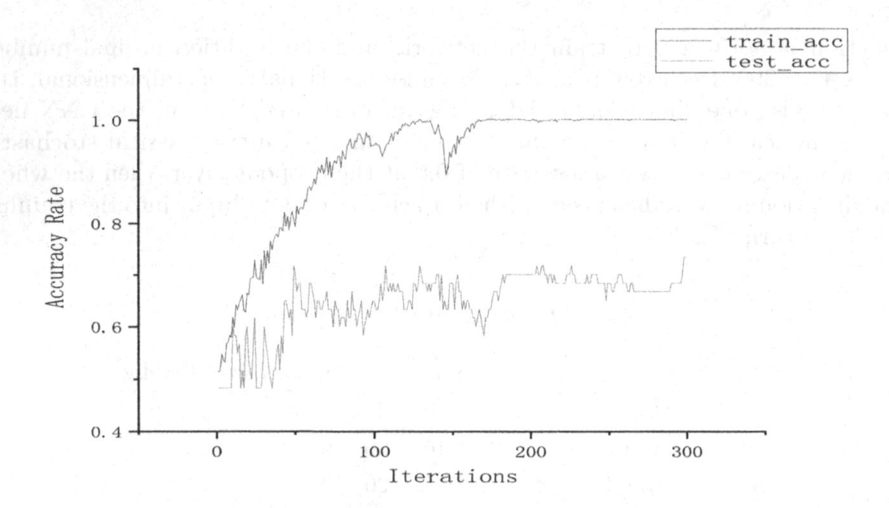

Fig. 4. The accuracy of train set VS test set after each iteration.

The method based on deep learning can definitely improve the accuracy of brain magnetic fields classification for motor imagery. A 1-dimensional convolutional network model is constructed, which can find spatial information rules of brain magnetic fields from sample data to realize automatic classification, and is suitable for the collected motor imagery data set. The model is simple and high accuracy, which meets the requirements of BCI system to a certain extent. Considering the shortcomings of the algorithm and the room for improvement of the experiment, the future work will further study the spatial information of brain magnetic fields and promote the development of BCI system.

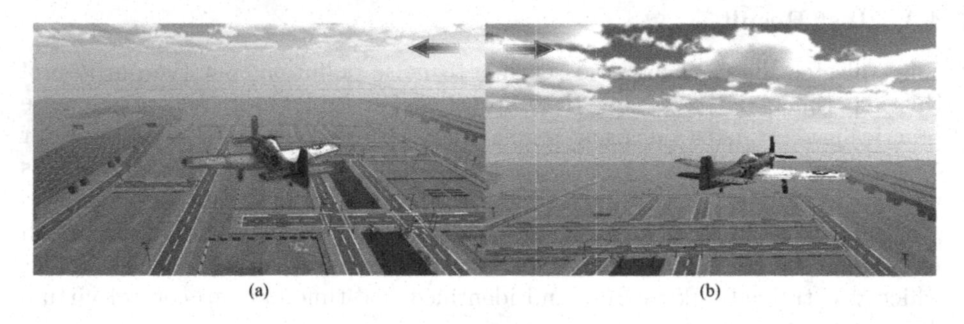

Fig. 5. Virtual drone is flighting in three dimensional space: (a) drone turn left. (b) drone turn right. All of operations is performing offline.

5 Conclusion and Future Work

In this study, we discussed the method for measuring brain's magnetic fields for motor imagery and further analyzed it's time-frequency feature. From that, enable flight movement for virtual drone in three-dimensional space controlled by three OPM channels data which was classified in machine learning. The focus of future work will be on optimizing experimental paradigm and comparing extensively performance of vary classification in order to improve the validity.

References

1. Pfurtscheller, G., Lopes da Silva, F.H.: Event-related EEG/MEG synchronization and desynchronization: basic principles. Clin. Neurophysiol. **110**(11), 1842–1857 (1999). https://doi.org/10.1016/s1388-2457(99)00141-8
2. Iivanainen, J., Zetter, R., Parkkonen, L.: Potential of on-scalp MEG: robust detection of human visual gamma-band responses. Hum. Brain Mapp. **41**(1), 150–161 (2020). https://doi.org/10.1002/hbm.24795
3. Kim, Y.J., Savukov, I.: Ultra-sensitive magnetic microscopy with an optically pumped magnetometer. Sci. Rep. **6**(1), 1–7 (2016). https://doi.org/10.1038/srep24773
4. Hill, R.M., Boto, E., Rea, M., Holmes, N., Leggett, J.: Multi-channel whole-head OPM-MEG: helmet design and a comparison with a conventional system. Neuroimage **219**(1), 116995 (2020)
5. Hämäläinen, M., Hari, R., Ilmoniemi, R.J., Knuutila, J., Lounasmaa, O.V.: Magnetoencephalography-theory, instrumentation, and applications to noninvasive studies of the working human brain. Rev. Mod. Phys. **65**(2), 413 (1993). https://doi.org/10.1103/RevModPhys.65.413
6. Osborne, J., Orton, J., Alem, O., Shah, V.: Fully integrated standalone zero field optically pumped magnetometer for biomagnetism. SPIE.Digital Library. California, United States (2018)
7. Pfurtscheller, G., Andrew, C.: Event-related changes of band power and coherence: methodology and interpretation. J. Clin. Neurophysiol. **16**(6), 512 (1999)
8. Leocani, L., Toro, C., Manganotti, P., Zhuang, P., Hallett, M.: Event-related coherence and event-related desynchronization/synchronization in the 10 Hz and 20 Hz EEG during self-paced movements. Electroencephalography and Clinical Neurophysiology/Evoked Potentials Section **104**(3), 199–206 (1997). https://doi.org/10.1016/S0168-5597(96)96051-7
9. Jingwei, L., Yin, C., Weidong, Z.: 2015 34th Chinese Control Conference (CCC). IEEE, Hangzhou, China (2015)
10. Jinzhen, L., Fangfang, Y., Hui, X.: Recognition of multi-class motor imagery EEG signals based on convolutional neural network. J. Zhejiang Univ. **55**(11), 2054–2066 (2021). https://doi.org/10.3785/j.issn.1008-973X.2021.11.005
11. Chunning, S., Yong, S., Zhenggao, N.: Deep learning-based method for recognition of motion imagery EEG signal. Transducer Microsyst. Technol. **41**(4), 125–128+133 (2022). https://doi.org/10.13873/j.1000-9787(2022)04-0125-04

Brain Controlled Manipulator System Based on Improved Target Detection and Augmented Reality Technology

Yiling Huang[1], Banghua Yang[1,2](✉), Zhaokun Wang[1], Yuan Yao[1], Mengdie Xu[1], and Xinxing Xia[1]

[1] School of Mechanical and Electrical Engineering and Automation, Research Center of Brain-Computer Engineering, Shanghai University, Shanghai 200444, China
yangbanghua@shu.edu.cn

[2] Engineering Research Center of Traditional Chinese Medicine Intelligent Rehabilitation, Ministry of Education, Shanghai 201203, China

Abstract. Aiming at the low detection and recognition rate of small and medium size targets in the current brain-controlled manipulator, this article proposes a method to solve the problem. Firstly, the self-made data set is expanded by using data enhancement technology and improving the robustness of the model, then improve the Faster-RCNN model by reducing the anchor boxes size, the mAP of the model has been increased by 2.62% on the original basis. After the improvement, combined with Augmented Reality (AR) technology to build a brain-controlled manipulator system. AR is used as a visual stimulator, The target position information is obtained through the improved target detection model, and the EEG signal of Steady-State Visual Evoked Potential (SSVEP) is recognized by Filter Bank Canonical Correlation Analysis (FBCCA), the grasping of the manipulator is controlled by decoding the EEG. 10 subjects participating in the grasping experiment, according to the experimental results, the grasping accuracy of the brain-controlled manipulator system is 92%, which verifies the effectiveness and portability of the system.

Keywords: BCI · Target detection · Augmented reality technology

1 Introduction

Brain-computer interface technology realizes the control of external equipment by establishing the connection between the brain and external equipment [1], which can help stroke patients or other people with limited mobility to improve their quality of life. A typical application is the brain-controlled robotic arm. Hortal E et al., based on Support Vector Machines (SVM) classification to control a robot arm [2], Chen X G et al., used steady-state visual to control the robotic arm [3], and Chen X et al., combined with eye tracker and motor imagination designed a brain-controlled robotic arm system [4].

The above research analysis shows that when the robot arm grabs the specified item, it needs to rely on the target detection algorithm to obtain the type and pixel coordinates,

X. Ying (Ed.): HBAI 2022, CCIS 1692, pp. 172–182, 2023.
https://doi.org/10.1007/978-981-19-8222-4_15

and then send the transformed coordinates to the robot arm for grabbing the item through coordinate transformation. In practical applications, in order to facilitate the grasping of the robotic arm, a small self-made data set is generally used, and the camera as the "eye" is generally placed on top of the grasped object. The target detection effect of the system is poor, and the grasping accuracy is low. And in the brain control system, the computer display is generally used as visual stimulation. Due to the setting of hardware, the convenience and practicability of the brain control system are insufficient.

Therefore, this paper proposes an improved algorithm to be applied to the brain-controlled manipulator system to improve recognition accuracy. Then, using FBCCA to analyze the brain EEG signal [5], the EEG signal is converted into a control signal to control the manipulator [6–8]. The system also integrates Augmented Reality technology, Augmented Reality device is used as a display to improve the portability of the SSVEP-BCI system.

2 Experimental and System Structure Design

2.1 Experimental Design

Select 10 healthy volunteers, including 8 males and 2 females, 20–26 years old. All volunteers' vision or corrected vision is normal. The SSVEP stimulus interface combined with Augmented Reality technology is shown in Fig. 1. The AR interface is used as the stimulation interface, and the flashing frequency of the square corresponds to the item one by one. According to the voice command, the subject needs to gaze at the flashing block of the corresponding object in the AR stimulus interface, the robotic arm is controlled to grasp the target. Each subject needs to perform 50 grasping experiments.

Fig. 1. SSVEP stimulation interface in AR.

The subjects who participated in the experiment be required to wear an EEG cap and AR device, then gaze at the corresponding object in the AR stimulation interface according to the voice command. After decoding by EEG, the EEG signal is converted into a control signal to control the robotic arm to grasp. Figure 2 shows the experimental scene.

Fig. 2. System experiment scene.

2.2 System Structure Design

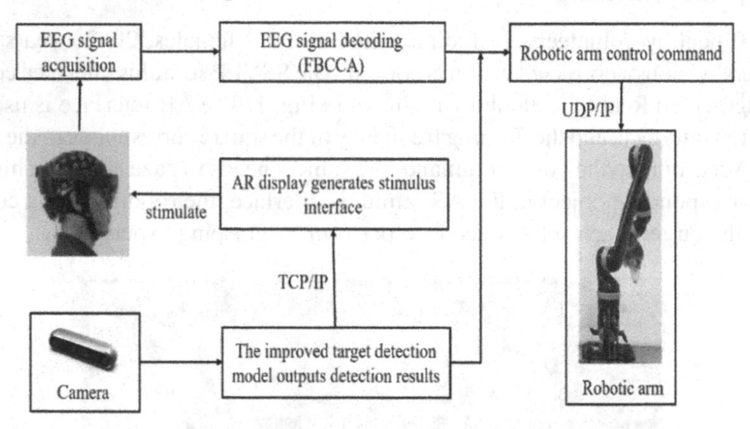

Fig. 3. System structure.

Figure 3 shows the structure of the brain-controlled manipulator system. BCI technology is based on SSVEP EEG signals. The system adopts Neuracle's 64-lead EEG cap to collect EEG signals, get real-time video stream by using Intel RealSense D435 camera, KINOVA J2S7S300 three-finger lightweight robotic arm for grasping. Which with redundant safety control and the control algorithm with singularity avoidance can be well satisfied with the brain control system. In order to improve the portability of the system, HoloLens glass (AR equipment) is used as a visual stimulator.

The whole system uses the improved detection algorithm for target detection, and the detection result will be sent to HoloLens glasses via TCP/IP communication protocol. Hololens glasses display visual stimulation interface, and then the FBCCA algorithm is used to decode EEG signals. Finally, the decoding result is converted into a control signal and sent to the robot arm through the UDP communication protocol, which controls the mechanical to grasp the target.

3 Methods

3.1 Improvement of Faster-RCNN

Figure 4 shows the basic Faster-RCNN network [9–11], The original images first go through the convolutional neural network for feature extraction. After the features are extracted by the convolutional layer will be divided and sent into the Region Proposal Networks (RPN) layer and the Region of Interesting (ROI) pooling layer. The RPN layer will select the possible candidate regions in the picture, and use it to assist the final target detection decision. ROI Pooling will collect the feature maps and proposals, extract the proposal feature maps after combining this information, then the fully connected layer will output the image category and location [12–14].

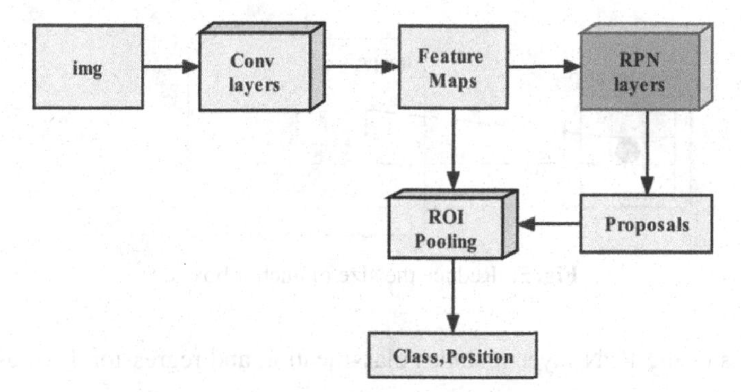

Fig. 4. Faster-RCNN network model structure.

The Faster-RCNN network model in this paper is improved by improving the RPN layer. Compared with several other deep learning-based target detection networks, Faster-RCNN target detection uses the RPN layer to generate candidate frames. By using the operation mode of the sliding window on Feature maps, the RPN layer changes the shape and size of the original image area corresponding to each point of the feature maps, to obtain anchor boxes of different shapes and sizes. In order to have wide coverage and ensure efficiency, three ratio changes of 1:2, 1:1 and 2:1 are adopted, and anchor boxes of three sizes of 128 * 128, 256 * 256 and 512 * 512 are used for transformation. In Faster-RCNN network training, anchor boxes divide positive and negative samples according to the IOU value between anchor boxes and ground truth boxes. If the IOU value between anchor boxes and any ground truth boxes is higher than 0.5, or it has the highest IOU relative to the value of ground truth, it is divided into positive samples, and small targets are easy to be divided into negative samples due to their small occupied area, so it will not participate in network training, this leads to the whole network fitting medium and large targets better.

Therefore, this System improves the anchor, so that more anchor boxes can be divided into positive samples. As shown in Fig. 5, the method reduces the size of three groups of anchor boxes to 64 * 64, 128*128 and 256*256 at the same proportion. The smaller anchor box can increase the IOU value of ground truth boxes, so as to improve the detection effect of small targets. Before building the model, we prepare the original self-made data set for model training and testing, the data set was enhanced by data processing to improve the robustness of the model, including flipping, rotating, brightness adjustment and adding Gaussian noise.

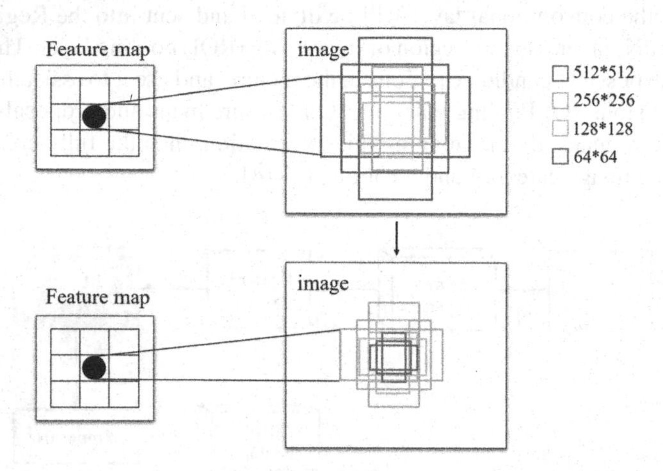

Fig. 5. Reduce the size of anchor boxes.

The loss of the RPN layer includes classification and regression loss, as shown in formula (1).

$$L(\{p_i\},\{t_i\}) = \frac{1}{N_{cls}}\sum_i L_{cls}(p_i,p_i^*) + \lambda\frac{1}{N_{reg}}\sum_i p_i^* L_{reg}(t_i,t_i^*) \tag{1}$$

In:

$$L_{cls}(p_i,p_i^*) = -\log[p_i^* p_i + (1-p_i^*)(1-p_i)] \tag{2}$$

$$L_{reg}(t_i,t_i^*) = R(t_i - t_i^*) \tag{3}$$

$$smooth_{L1} = \begin{cases} 0.5x^2 & if\ \ |x| < 1 \\ |x| - 0.5 & otherwise \end{cases} \tag{4}$$

p_i represents the predicted probability; p_i^* is 1 or 0. t_i represents the bounding box regression parameter; t_i^* represents coordinates of the ground truth box; N_{cls} represents the total number of samples in a mini-batch; N_{reg} represents the number of anchor locations; λ is the balance weight; R is $smooth_{L1}$ function.

3.2 AR Technology Generates Stimulation Interface

The system uses HoloLens glass to replace the traditional computer screen as the stimulation display, so as to get rid of the limitation of the equipment and improve the portability of the system. The stimulation display requires a three-dimensional square flashing at a specific flashing frequency, and the whole stimulation interface is built based on AR technology [15]. First, use 3D MAX software to make three-dimensional blocks, and then add the C# script to the blocks in unity3d software. Use sinusoidal coding to generate stimulation sequence through formula (5), and adjust the transparency of the screen according to the sequence to achieve specific flicker frequency.

$$stim(n, f, \Phi) = 1/2\{1 + sin[2\pi f(n/R)] + \Phi\} \quad (5)$$

sin stands for sinusoidal signal; R is the refresh rate of the display and the sampling rate of the sinusoidal signal; n is the sequence number of each frame of the display in the stimulation sequence; In the settings of HSG rendering, usually 0 means opaque and 1 means fully transparent, Therefore, $stim(n, f, \Phi)$ is the value range of transparency $(0 - 1)$. Theoretically, any frequency with any phase and less than half of the display refresh rate can be realized by sampling the sinusoidal coding paradigm.

After that, the AR device receives the information from the target detection module through TCP/IP communication protocol to display the visual stimulation interface. The flow chart of stimulation interface generation is shown in Fig. 6.

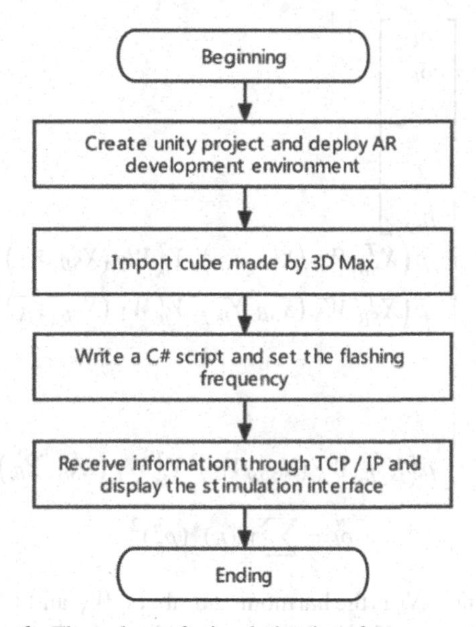

Fig. 6. Flow chart of stimulation interface generation.

3.3 EEG Signal Analysis

SSVEP refers to the continuous response of the human visual cortex to a fixed frequency of visual stimulation, characterized by a high transmission rate with little or no training. This system uses the FBCCA algorithm to decode the SSVEP signal, and the decoded information is converted into a control signal to control the robotic arm. FBCCA is an improvement of the CCA algorithm, based on CCA, a filter bank is added to make use of the EEG signal harmonic components that are not fully utilized in the traditional CCA algorithm, and improve the accuracy of the algorithm [Ge et al., 2019]. Firstly, N bandpass filters are constructed, then read the EEG data according to the cycle, and obtain N bandpass filtering results $X_{SB1} \ldots X_{SBN}$. a series of CCA methods are used to calculate the correlation between reference signal Y_{fk} and EEG signal, as shown in formula (7). Vector ρ_k represents the corresponding correlation coefficient value of the kth reference signal, which contains the corresponding correlation coefficient value of the N subbands. Then, by the formula (8), the N subbands components in ρ_k do a weighted sum of squares. After calculating all $\widetilde{\rho_k}$, a maximum $\widetilde{\rho_k}$ can be determined, and the corresponding frequency is the final result [16, 17].

$$Y_{fk} = \begin{bmatrix} \sin(2\pi f_k t) \\ \cos(2\pi f_k t) \\ \cdots\cdots \\ \sin(2\pi N_h f_k t) \\ \cos(2\pi N_h f_k t) \end{bmatrix} \tag{6}$$

$$\rho_k = \begin{bmatrix} \rho_{k_1} \\ \rho_{k_2} \\ \cdot \\ \cdot \\ \cdot \\ \rho_{k_N} \end{bmatrix} = \begin{bmatrix} \rho\left(X_{SB_1}^T W_X\left(X_{SB_1} Y_{fk}\right), Y_{fk}^T W_Y\left(X_{SB_1} Y_{fk}\right)\right) \\ \rho\left(X_{SB_2}^T W_X\left(X_{SB_2} Y_{fk}\right), Y_{fk}^T W_Y\left(X_{SB_2} Y_{fk}\right)\right) \\ \cdot \\ \cdot \\ \cdot \\ \rho\left(X_{SB_N}^T W_X\left(X_{SB_N} Y_{fk}\right), Y_{fk}^T W_Y\left(X_{SB_N} Y_{fk}\right)\right) \end{bmatrix} \tag{7}$$

$$\widetilde{\rho_k} = \sum w(n)^* (\rho_k^n)^2 \tag{8}$$

f_k is stimulation frequency, N_h is the harmonic numbers, W_X and W_Y are CCA correlation matrices; $w(n) = n^{-a} + b$, $n \epsilon [1, N]$, Where, a and b are both constants and their values are determined by the time when the classifier performance reaches its best case.

4 Experimental results

4.1 Target Detection Model Test

The self-made dataset used in the experiment included 10 common objects in daily life. Firstly, through data enhancement, the data set was expanded to 4000 pieces. Then, the improved Faster-RCNN model, which reduced the size of anchor boxes, was used for training and testing. Table 1 shows model parameters, and Table 2 shows the test results.

Table 1. Parameter setting

Parameter name	Parameter value
Iterations	50000
Learning rate	0.001
Batch size	128
Momentum	0.9
Gamma	0.1

Table 2. Mean average precision of different target detection models

Target detection model	Number of data	mAP (%)
Faster-RCNN	4000	90.02
Improved model	4000	92.64

The results indicate the mean average precision (mAP) of the method is increased by 2.62% than before the improvement. It shows that reduce anchor box size can effectively improve the detection accuracy of small targets.

4.2 EEG Recognition Results Using AR

EEG signals with 4 harmonics and 3 s of signal acquisition time are selected, using FBCCA to identify the brain EEG signal. Test the recognition accuracy when AR is used as a visual stimulator. The subjects looked at the flashing blocks displayed in AR, each subject looked at objects in turn as a trail, and 10 subjects completed five trails respectively. According to the results shown in Fig. 7, the accuracy of ten subjects can reach 90% and above, and the average recognition accuracy can reach 92.6%. Experiments show that when AR technology is integrated and the FBCCA algorithm is used for EEG signal recognition has a good effect.

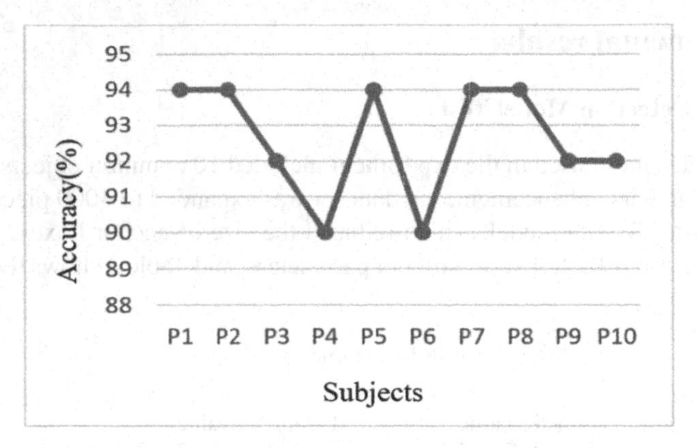

Fig. 7. The recognition accuracy rate of EEG signal.

4.3 System Test

The improved Faster-RCNN model was used to train the enhanced data set, and the target detection model obtained was used for real-time target recognition. Then, The AR device is used as a visual stimulator, and the EEG signal is decoded and transformed into a control signal through the FBCCA algorithm to control the grasping of the manipulator. In the experiment, the accuracy of target recognition, target detection speed and the recognition accuracy of the SSVEP EEG signals of 10 subjects were recorded. The brain-controlled manipulator system combined with the augmented reality test results is shown in Table 3.

Table 3. Brain-controlled manipulator system test results.

Subject	FPS	Target detection accuracy (%)	SSVEP recognition accuracy (%)	Grasping accuracy (%)
P1	21.1	96	92	92
P2	21.3	94	90	90
P3	20.8	96	96	96
P4	21.2	96	94	92
P5	21.2	92	92	92
P6	20.8	98	92	92
P7	20.9	96	94	92
P8	21.1	94	90	90
P9	21.3	92	92	92

(continued)

Table 3. (*continued*)

Subject	FPS	Target detection accuracy (%)	SSVEP recognition accuracy (%)	Grasping accuracy (%)
P10	21.0	94	92	92
Mean	21.07	94.8	92.4	92

According to the experimental results, the target detection accuracy of improved Faster-RCNN target detection is 94.8%, and the detection speed has not significantly reduced compared with before the improvement. The EEG recognition accuracy and grasping accuracy of the system are 92.4% and 92.0% respectively. Experiments show the feasibility of the system.

5 Concludes

In this paper, the self-made data set was enhanced to expand the data set and improve the model generalization ability. Through a way to improve the Faster-RCNN network, which improved the target detection accuracy by 2.62%. And on this basis, combined with Augmented Reality technology, construct a brain-controlled manipulator system based on SSVEP, the accuracy of EEG recognition and robotic grasping accuracy can reach 92.4% and 92%, respectively, showing good recognition and grasping effect. Through augmented reality technology, AR equipment is used to replace computer monitors, which improves the convenience and practicability of the system. It is suitable for medical rehabilitation and other fields to help the elderly and patients with movement disorders assist daily life, and improve the quality of their life. It is also of great significance to promote the practical application of brain-computer interface technology.

Acknowledgements. This project is supported by the National Natural Science Foundation of China (No. 61976133), Major scientific and technological innovation projects of Shan Dong Province (No. 2019JZZY021010), Shanghai Industrial Collaborative Technology Innovation Project (No. 2021-cyxt1-kj14), National Defense Basic Scientific Research Program of China (JCKY2021413B002).

References

1. Wang, Y., Chen, X., Gao, X., et al.: A benchmark dataset for SSVEP-based brain–computer interfaces. IEEE Trans. Neural Syst. Rehabil. Eng. **25**(10), 1746–1752 (2017)
2. Hortal, E., Planelles, D., Costa, A., et al.: SVM-based brain-machine interface for controlling a robot arm through four mental tasks. Neurocomputing **151**, 116–121 (2015)
3. Chen, X.G., Zhao, B., Liu, M., et al.: Design and implementation of controlling robotic arms using steady-state visual evoked potential brain-computer interface. Biomedical Engineering and Clinical Medicine (2018)

4. Chen, X., Zhao, B., Wang, Y., et al.: Control of a 7-DOF robotic arm system with an SSVEP-based BCI. Int. J. Neural Syst. **28**(8), 1850018 (2018)
5. Ge, S., Jiang, Y., Wang, P., et al.: Training-free steady-state visual evoked potential brain-computer interface based on filter bank canonical correlation analysis and spatiotemporal beamforming decoding. IEEE Trans. Neural Syst. Rehabil. Eng. **27**(9), 1714–1723 (2019)
6. Zhang, R., Li, Y., Yan, Y., et al.: Control of a wheelchair in an indoor environment based on a brain-computer interface and automated navigation. IEEE Trans. Neural Syst. Rehabil. Eng. **24**(1), 128–139 (2016)
7. Parmar, P., Joshi, V.G.A., Joshi, A.: Brain-computer interface: a review. In: Nirma University International Conference on Engineering. IEEE (2016)
8. Jeong, J.H., Shim, K.H., Kim, D.J., et al.: Brain-controlled robotic arm system based on multi-directional CNN-BiLSTM network using EEG signals. IEEE Trans. Neural Syst. Rehabil. Eng. **28**(5), 1226–1238 (2020)
9. Li, L., Zeng, P.: Apple target detection based on improved faster-RCNN framework of deep learning. Mach. Des. Res. **35**, 24–27 (2019)
10. Fan, T.: Research and realization of video target detection system based on deep learning. Int. J. Wavelets Multiresolut. Inf. Process. **18**(1), 280–292 (2020)
11. Lv, S., Liu, K., Qiao, Y., et al.: Automatic defect detection based on improved faster RCNN for substation equipment. J. Phys. Conf. Ser. **1544**, 012157 (2020)
12. Harianto, R.A., Pranoto, Y.M., Gunawan, T.P.: Data augmentation and faster RCNN improve vehicle detection and recognition. In: 2021 3rd East Indonesia Conference on Computer and Information Technology (2021)
13. Jiang, D., Li, G., Tan, C., et al.: Semantic segmentation for multiscale target based on object recognition using the improved Faster-RCNN model. Future Gener. Comput. Syst. **123**(1), 94–104 (2021)
14. Mo, N., Yan, L.: Improved faster RCNN based on feature amplification and oversampling data augmentation for oriented vehicle detection in aerial images. Remote Sensing **12**(16), 2558 (2020). https://doi.org/10.3390/rs12162558
15. Ke, Y., Liu, P., An, X., Song, X., Ming, D.: An online SSVEP-BCI system in an optical see-through augmented reality environment. J. Neural Eng. **17**(1), 016066 (2020). https://doi.org/10.1088/1741-2552/ab4dc6
16. Shi, N., Wang, L., Chen, Y., et al.: Steady-state visual evoked potential (SSVEP)-based brain–computer interface (BCI) of Chinese speller for a patient with amyotrophic lateral sclerosis: a case report. J. Neurorestoratology **8**(1), 40–52 (2020)
17. Lin, C.T., Chiu, C.Y., Singh, A.K., et al.: A wireless multifunctional SSVEP-based brain-computer interface assistive system. IEEE Trans. Cogn. Dev. Syst. **11**(3), 375–383 (2019)

Optimization of Stimulus Color for SSVEP-Based Brain-Computer Interfaces in Mixed Reality

Feng He[1,2], Jieyu Wu[1], Xiaolin Xiao[1,2](✉), Runyuan Gao[1], Weibo Yi[3], Yuanfang Chen[3], Minpeng Xu[1,2], Tzyy-Ping Jung[4], and Dong Ming[1,2]

[1] College of Precision Instruments and Optoelectronics Engineering, Tianjin University, Tianjin 300072, China
xiaoxiao0@tju.edu.cn

[2] Academy of Medical Engineering and Translational Medicine, Tianjin University, Tianjin 300072, China

[3] Beijing Institute of Mechanical Equipment, Beijing 100143, China

[4] Swartz Center for Computational Neuroscience, University of California, San Diego, CA 92093, USA

Abstract. Visual Brain-Computer Interfaces (BCIs) often use LCDs or LEDs to present flickering stimuli, which limits its application scenarios. Mixed reality head-mounted displays (MRHMDs) have the potential of improving the practical applications for BCIs. However, it's unclear whether the visual stimulus color designed for traditional BCIs still works in mixed reality. Therefore, this study developed a 10-command SSVEP-BCI system in mixed reality using Hololens2, and explored the system's performance with two stimulus colors (white and red) against four monochrome backgrounds (green, blue, white, and black). Eight subjects participated in the experiment. Cross-correlation task-related component analysis (xTRCA) was used to recognize the target command. Results showed that both stimulus and background colors affect BCI performance. Specifically, the white stimulus color significantly outperformed the red one on the blue and black backgrounds, while the red stimulus outperformed the white one on the green and white backgrounds. The color contrast ratio (CCR) between the background and stimulus colors correlated positively with SSVEP recognition accuracy. In mixed reality, SSVEP-based BCIs must optimize visual stimulus color, and the CCR is an important criterion for optimization.

Keywords: Steady-state visual evoked potential (SSVEP) · Mixed reality (MR) · Color contrast ratio (CCR) · Cross-correlation task-related component analysis (xTRCA)

1 Introduction

1.1 A Subsection Sample

Brain-Computer Interfaces (BCIs) provide a direct communication channel between the human brain and the outside world [1–3]. Electroencephalography (EEG)-based

X. Ying (Ed.): HBAI 2022, CCIS 1692, pp. 183–191, 2023.
https://doi.org/10.1007/978-981-19-8222-4_16

BCIs have been widely used and achieved high system performance [4]. Among various BCI paradigms, Steady-State Visual Evoked Potentials (SSVEPs) have become one of the most used EEG signals for BCIs due to their high signal-to-noise rate (SNR) and information transfer rate (ITR) [5].

Most SSVEP-BCIs present stimuli on an LCD screen or LED array to elicit the time- and phase-locked SSVEPs [6]. However, LCD and LED displays were inconvenient to use in the real-world environment, hindering BCI applications. In recent years, some studies demonstrated the feasibility of implementing SSVEP-BCIs in an augmented reality (AR) or mixed reality (MR) environment using head-mounted displays (HMDs). Specifically, flickering stimulus could be presented on HMDs, and be superimposed on the surrounding environment. It provided BCIs' a new way to interact with the real world. Takano et al. [7] developed a four-command SSVEP-BCI in AR in 2011. Their system reached an average classification accuracy of about 82%, and controlled the light successfully. In 2020, Jaehoon et al. [8] used HMDs to build a motor imagery (MI)-based BCI and an SSVEP-based BCI to control a quadcopter, which achieved an average accuracy of 52.50% and 72.40%, respectively. In addition, several studies also validated the feasibility and practicability of BCIs based on HMDs [9–11].

Previous SSVEP research has explored the use of colored visual stimuli. Tello et al. [12] reported that red stimuli elicited a larger SSVEP amplitude than blue and green stimuli. The reason for this, they stated, was that the red color attracted greater attention. However, Cao et al. [13] claimed that white stimuli were better than blue and green ones because they can activate all three kinds of cone cells in human eyes. Colored SSVEP stimuli have been the subject of many studies, however little research has been conducted in augmented or mixed reality environments, specifically when it comes to the interaction between stimulus and background colors.

This study used Microsoft Hololens2 to create an SSVEP-BCI system and tested its performance using two stimulus colors and four backdrop colors. Hololens2, being one of the most advanced HMDs, can display holographic pictures of stimuli onto the real environment, giving BCI users a more engaging experience. Color contrast ratios (CCRs) were used to evaluate the color attribute, and cross-correlation task-related component analysis (xTRCA) was used to recognize SSVEPs due to its superiority in mitigating phase jitters across trials caused by HMDs [14].

2 Methods and Materials

2.1 Experimental Protocol

The Hololens2, a Microsoft MR-HMD device, was used in this investigation. The subjects wore the Hololens2 while wearing the EEG recording cap in the experiment. In the experiment, the subjects wore the Hololens2 while wearing the EEG recording cap. The stimulation applications were created using the Unity 3D engine and be build on Universal Windows Platform (UWP) to establish a C# project. Then the application will be deployed to Hololens2 via USB. Event triggers were generated in Hololens2 at the beginning of stimulation presentation and transmitted to the EEG amplifier via the user datagram protocol (UDP). These triggers were used to extract data epochs, which were then examined in MATLAB.

The refresh rate of Hololens2 is 60 Hz. The sampled sinusoidal stimulation method [15] was used to encode 10 commands. The frequency range was 8 to 12.5 Hz with a step of 0.5 Hz. The stimuli were spread around the central visual region such that the participants' central visual field was not occluded. Figure 1 shows a schematic representation of the visual stimulus layout. Each stimulus was rendered within a 6° × 6° square in the visual angle. Hololens2 projected the stimulation plane 75 cm in front of the subject. Each trial started with a cue (a white cross " +") that located in the center of the stimuli to indicate the next target. The cue lasted 1 s and the subjects were asked to shift their focus to the target during that time. After that, the stimuli flickered simultaneously for 1.5 s.

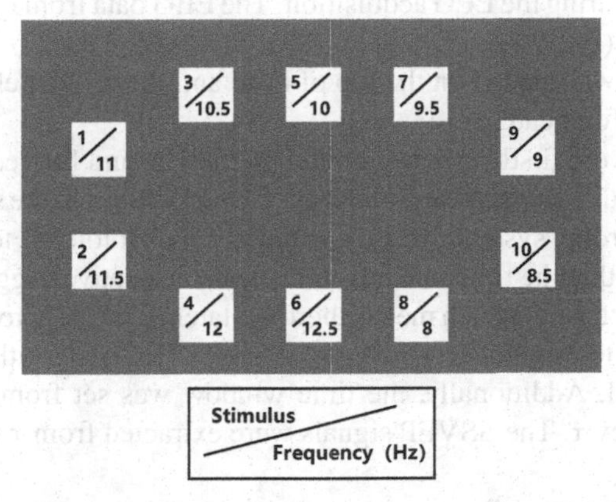

Fig. 1. SSVEP-BCI stimulation interface based on Hololens2.

This study consisted of eight experimental conditions created by combining two stimulus colors: white (RGB value: [255 255 255]), red (RGB value: [255 0 0]) with four background colors: green (RGB value: [0 255 0]), blue (RGB value: [0 0 255]), white (RGB value: [255 255 255]) and black (RGB value: [0 0 0]). All of the background colors are common in everyday life. We hung a large monochrome poster on the wall to simulate a monochrome background. The background color in the stimulus program is transparent, so that the observation of the monochrome poster will not be affected. The participants were asked to adjust the angle of the projection so that the stimuli were presented in the center of the poster. Before the experiment, subjects were told to focus on the stimuli during the stimulation period and try their best to avoid blinking. Six trials were acquired for each target, resulting in a 150-s bout for each condition. Between the two consecutive experiment conditions, the subjects would take a short break to avoid visual fatigue. The system's performance with the same stimuli on different color backgrounds was investigated to determine the optimal stimulus color in relation to the background color.

2.2 Participants

A total of eight healthy subjects (4 males and 4 females, 19 to 23 years old) whose vision were normal or corrected participated in this study. It's noted that none of them was color-blind. They had read and signed informed consents that were approved by the Research Ethics Committee of Tianjin University before the experiment.

2.3 EEG Acquisition and Data Pre-processing

This study used a Neuroscan Synamps2 system to acquire EEG data. The sampling rate of the amplifier was set to 1000 Hz. A 50 Hz notch filter and a band-pass filter (0.1 – 100 Hz) were used during the EEG acquisition. The EEG data from nine channels around the occipital region (PZ, POZ, PO3, PO4, PO5, PO6, OZ, O1, O2) were collected. The reference electrode was placed on the top of head and the ground electrode was placed in the center of the forehead.

The EEG data were firstly down-sampled to 250 Hz, and filtered with a Chebyshev filter to 6–15 Hz. In this study, the event triggers, which calibrate the start time of SSVEP signals in EEG recording system, were transmitted from Hololens2 to the EEG amplifier via UDP. We found that the time intervals between consecutive triggers recorded by the EEG amplifier are jittering, which means there are latency jitters across trials which were extracted according to original event triggers. Using xTRCA algorithm can optimize the latency of each trial. Additionally, the time window was set from −0.2 s to 1.7 s to determine the latency τ. The SSVEP signals were extracted from τ to $\tau + 1.5$ s.

2.4 Classification Algorithm

This study used xTRCA to compensate phase jitter across trials and pattern recognition [17]. Based on Task-Related Component Analysis (TRCA) [16], xTRCA constructs a linear spatial filter and then optimizes trial timings of single trials using trial reproducibility as an objective function. The temporal optimization compensates trial-by-trial latency variability in SSVEPs.

Considering a set of trials $\boldsymbol{X} \in R^{N \times T \times K}$, we suppose that the timing vector formed by the initial phase of trials is $\boldsymbol{t}_0 = [\tau_0^1, \ldots, \tau_0^k, \ldots, \tau_0^K]$, where N is the number of channels, T is number of time points (includes a complete trial and the fragments before and after it), K is number of trials, and τ_0^k is the assumed initial phase of $k - th$ trial $\boldsymbol{X}^k$. According to TRCA in [15], we can solve a Rayleigh-Ritz eigenvalue problem as follows

$$\widehat{\boldsymbol{w}} = \underset{\boldsymbol{w}}{\operatorname{argmax}} \frac{\boldsymbol{w}^T \boldsymbol{S} \boldsymbol{w}}{\boldsymbol{w}^T \boldsymbol{Q} \boldsymbol{w}} \tag{1}$$

where $\boldsymbol{S}$ is a covariance matrix between all different two trials

$$\boldsymbol{S} = \frac{1}{K(K-1)T_0} \sum\nolimits_{k \neq l}^{K} \boldsymbol{X}^k \left(\boldsymbol{X}^l\right)^T \in R^{N \times N} \tag{2}$$

and $\boldsymbol{Q}$ is calculated with a concatenated matrix of all the trials $\boldsymbol{X}$

$$\boldsymbol{Q} = \frac{1}{TK}\sum\nolimits_{k,j=1}^{N} \boldsymbol{X}_k\left(\boldsymbol{X}_j\right)^T \tag{3}$$

After we got a $\boldsymbol{w}$, we can optimize the timing vector $\boldsymbol{t}$

$$\hat{t} = \underset{t}{\operatorname{argmax}} \frac{\boldsymbol{w}^T \boldsymbol{S}(\boldsymbol{t})\boldsymbol{w}}{\boldsymbol{w}^T \boldsymbol{Q}\boldsymbol{w}} \tag{4}$$

The timing vector should be optimized to maximize the eigenvalue, which is a measure of reproducibility and hence also serves as the objective function for this cross-correlation solution. As a result, we can calculate the cross-correlation coefficient between the $k-th$ trial and other trials to determine the time shifting, $\boldsymbol{w}$ and $\boldsymbol{t}$ would be repeatedly optimized until the dominant eigenvalue converges within a specified threshold.

After phase compensation, the canonical correlation analysis (CCA) was used to recognize the patterns of SSVEPs. The detailed descriptions of TRCA and xTRCA can be found in [16] and [17] respectively.

2.5 Calculation of Color Contrast

Color contrast ratios (CCRs) were used to evaluate the visual differences between stimulus color and background colors in this study. The calculation method was based on the definition of color contrast in Web content accessibility guidelines (WCAG) [18]. Supposed that there are two colors, and the RGB values are: $C_1 : [R_1, G_1, B_1]$ and $C_2 : [R_2, G_2, B_2]$ (the RGB value range from 0 to 1). The CCR between C_1 and C_2 is calculated as

$$CCR = \frac{\max(L_1, L_2) + 0.05}{\min(L_1, L_2) + 0.05} \tag{5}$$

where L is the relative luminance of the colors, which is calculated as

$$L = 0.2126R' + 0.7152G' + 0.0722B' \tag{6}$$

where R', G', B' are defined as

$$R' = \begin{cases} \frac{R}{12.92}, R \le 0.03928 \\ \left(\frac{R+0.055}{1.055}\right)^{2.4}, R > 0.03928 \end{cases} \tag{7}$$

G' and B' are calculated in the same way. More details of this process could be referred to [18].

3 Result and Analysis

Firstly, we investigated the characteristics of SSVEPs induced by different stimuli. Figure 2 shows a sample SSVEP elicited by white and red stimuli flashed at 8 Hz

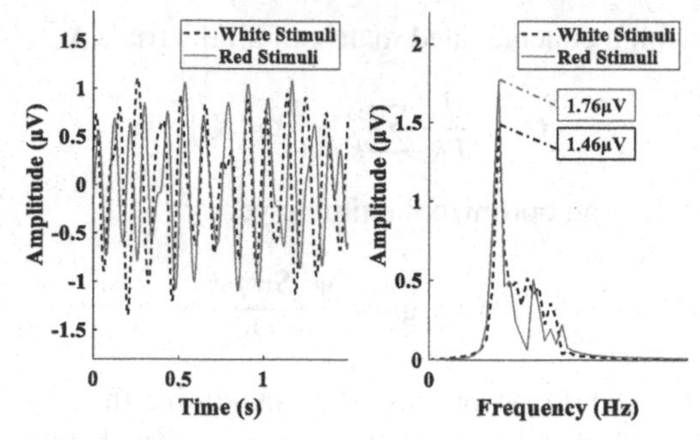

Fig. 2. The temporal waveform and amplitude spectrum of 8 Hz SSVEPs, under the green background.

on a green background across all subjects. The spatial filter obtained by xTRCA was used to weight the nine channels. We found that the amplitude of the target frequency was lower for the white stimuli than for the red ones. The results showed that different stimulus colors elicited different SSVEP responses.

We then calculated the classification accuracies for all eight conditions. As a result, the classification accuracies were significantly different among the eight experimental conditions. Specifically, the accuracy was 87.71 ± 6.00% for green background and white stimuli, 93.33 ± 3.54% for green background and red stimuli, 92.50 ± 5.25% for blue background and white stimuli, 87.92 ± 5.39% for blue background and red stimuli, 88.96 ± 5.77% for white background and white stimuli, 91.67 ± 5.14% for white background and red stimuli, 94.58 ± 3.31% for black background and white stimuli, 88.75 ± 5.25% for black background and red stimuli, respectively. In conclusion, the best stimulus colors on the green, blue, white and black backgrounds were red, white, red and white, respectively.

Two-way (stimuli colors and background colors) repeated-measure ANOVAs were performed on the accuracies. The results showed that there was significant interaction between stimuli and background colors ($F(3,21) = 9.234$; $p < 0.001$***). Figure 3 and Fig. 4 show the results of pairwise comparisons, which used paired t-tests. As shown in Fig. 3, paired t-tests found that white stimuli significantly outperformed the red ones on the blue ($p < 0.05$*) and black backgrounds ($p < 0.05$*). They had no significant difference for the white and green background. Figure 4 indicate the significant difference between each pair of the four backgrounds by paired t-tests. For white stimuli, the black background ($p < 0.05$*) and blue background ($p < 0.05$*) was significantly different from the green background. For red stimuli, the green background was significantly different from the blue background ($p < 0.05$*).

In 2012, Conway et al. [19] explored the neural basis for spatial color contrast and temporal color contrast in the primary visual cortex (V1) of the alert macaque. They found double-opponent color cells that were sensitive to opposite colors (red and green) in the V1 cortex and responded strongly when the macaque saw red stimuli surrounded by green (spatial color contrast). In this study, the CCRs of colors between the backgrounds and stimuli were calculated to assess the correlation between the color contrast and performance of SSVEP-BCIs. Figure 5 shows the correlation between CCR and accuracy.

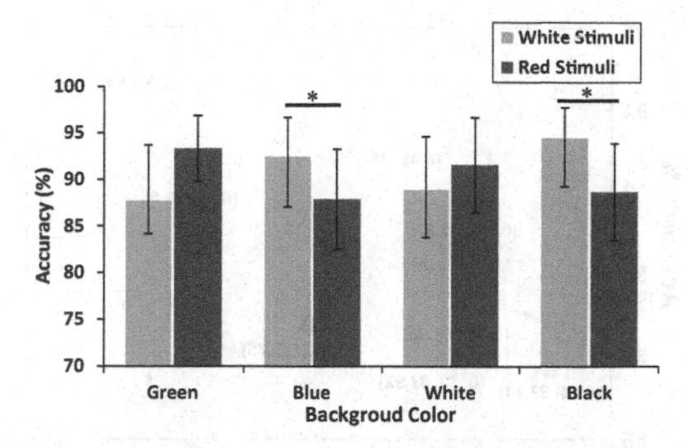

Fig. 3. The average accuracy of 8 experiments across 8 subjects and significant difference between two kinds of stimuli on four backgrounds obtained by one-way ANOVAs, (* $p < 0.05$).

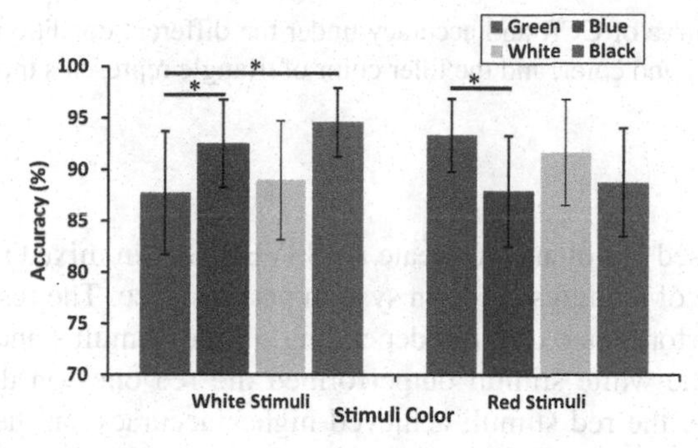

Fig. 4. The significant differences between four backgrounds with the same stimuli obtained by one-way ANOVAs (* $p < 0.05$).

The x-axis is the logarithm of CCR between background and stimulus colors, while the y-axis represents the accuracy under different conditions. The Pearson correlation coefficient between CCR and accuracy was 0.68 (0.69 between log(CCR) and accuracy). CCR was positively correlated with SSVEP recognition accuracy. Specifically, the higher the CCR between background and stimuli, the better the accuracy. It's worth noting that the Hololens2 has a transmittance of 0.85. As a result, the white background seen by the subjects appears to be gray. The CCR between gray and red is 1.96 ($\log(1.96) = 0.29$), whereas gray and white have a CCR of 2.03 ($\log(2.03) = 0.31$), which is very comparable. This might explain why there was no significant difference between white and red stimuli on a white background as shown in Fig. 3.

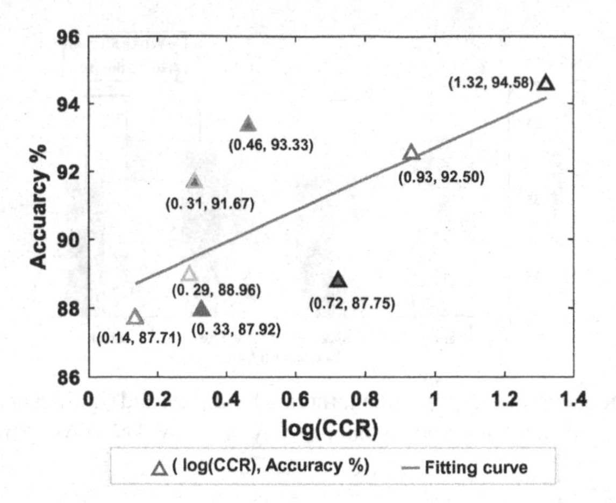

Fig. 5. The fitting curve of CCR and accuracy under the different conditions. The line color of triangle is the background color, and the filler color of triangle represents the stimulus color.

4 Conclusion

In this study, we used Hololens2 to create an SSVEP-BCI in mixed reality and investigated the influence of stimulus color on system performance. The results demonstrated that the system performance differed depending on the stimulus and background colors. Specifically, the white stimuli outperformed the red ones on the blue and black backgrounds, while the red stimuli achieved higher accuracy on the green and white backgrounds. Notably, the CCR was introduced to explain these differences and we found that CCR was positively correlated with SSVEP recognition accuracy. As a result, in mixed reality, BCIs must optimize visual stimulus color against various backgrounds.

Acknowledgments. Research supported by National Natural Science Foundation of China (No. 62106170, 62122059, 81925020, 61976152), Introduce Innovative Teams of 2021 "New High School 20 Items" Project (2021GXRC071), and Tianjin Key Technology R&D Program (No. 17ZXRGGX00020).

References

1. Wolpaw, J.R., Birbaumer, N., McFarland, D.J., Pfurtscheller, G., Vaughan, T.M.: Brain-computer interfaces for communication and control. Clin. Neurophysiol. **113**(6), 767–791 (2016)
2. Xiao, X., et al.: Enhancement for P300-speller classification using multi-window discriminative canonical pattern matching. J. Neural Eng. **18**(4), 046–079 (2021)
3. Wolpaw, J.R.: Brain-computer interfaces as new brain output pathways. J. Physiol. **579**(3), 613–619 (2007)
4. Ramadan, R., Vasilakos, A.: Brain computer interface: control signals review. Neurocomputing **27**(5), 26–44 (2017)
5. Wang, Y., Gao, X., Hong, B., Jia, C., Gao, S.: Brain-Computer Interfaces Based on Visual Evoked Potentials. IEEE Eng. Med. Biol. Mag. **223**(5), 64–71 (2008)

6. Wu, Z., Lai, Y., Xia, Y., Wu, D., Yao, D.: Stimulator selection in SSVEP-based BCI. Med. Eng. Phys. **30**(8), 1079–1088 (2008)
7. Kouji, T., Naoki, H., Kenji, K.: Towards intelligent environments: an augmented reality–brain–machine interface operated with a see-through head-mount display. Front. Neurosci. **5**, 60 (2011)
8. Choi, J., Jo, S.: Application of hybrid brain-computer interface with augmented reality on quadcopter control. In: 8th International Winter Conference on Brain-Computer Interface (BCI), pp. 1–5. IEEE, Gangwon, Korea (South) (2020)
9. Angrisani, L., Arpaia, P., Moccaldi, N., Esposito, N.: Wearable augmented reality and brain computer interface to improve human-robot interactions in smart industry: a feasibility study for SSVEP signals. In: 4th IEEE International Forum on Research and Technology for Society and Industry, pp. 1–5. IEEE, Palermo, Italy (2018)
10. Horii, S., Nakauchi, S., Kitazaki, M.: AR-SSVEP for brain-machine interface: estimating user's gaze in head-mounted display with USB camera. In: 2015 IEEE Virtual Reality, pp. 193–194. IEEE, Arles, France (2015)
11. Meng, W., Li, R., Zhang, R., Li, G., Zhang, D.: A wearable SSVEP-based BCI system for quadcopter control using head-mounted device. IEEE Access **6**, 26789–26798 (2018)
12. Tello, R.J.M.G., Müller, S.M.T., Ferreira, A., Bastos, T.F.: Comparison of the influence of stimuli color on steady-state visual evoked potentials. Res. Biomed. Eng. **1**(3), 1041–1047 (2015)
13. Cao, T., Wan, F., Mak, P.U., Mak, P.I., Vai, M.I., Hu, Y.: Flashing color on the performance of SSVEP-based brain-computer interfaces. In: Annual International Conference of the IEEE Engineering in Medicine and Biology Society, pp. 1819–1822. IEEE, San Diego, CA, USA (2012)
14. Ke, Y., Liu, P., An, X., Song, X., Ming, D.: An online SSVEP-BCI system in an optical see-through augmented reality environment. J. Neural Eng. **17**(1), 016066 (2020)
15. Wang, Y.T., Jung, T.P.: Visual stimulus design for high-rate SSVEP BCI. Electron. Lett. **46**(15), 1057–1058 (2010)
16. Nakanishi, M., Wang, Y., Chen, X., Wang, Y.T., Gao, X., Jung, T.P.: Enhancing detection of SSVEPs for a high-speed brain speller using task-related component analysis. IEEE Trans. Biomed. Eng. **65**(1), 104–112 (2017)
17. Tanaka, H., Miyakoshi, M.: Cross-correlation task-related component analysis (xTRCA) for enhancing evoked and induced responses of event-related potentials. Neuroimage **197**, 177–190 (2019)
18. Consortium, W.: Web content accessibility guidelines (WCAG) 2.0 (2015)
19. Conway, B.R., Hubel, D.H., Livingstone, M.S.: Color contrast in macaque V1. Cereb. Cortex **12**(9), 915–925 (2002)

Brain Related Research

White Matter Maturation and Hemispheric Asymmetry During Childhood Based on Chinese Population

Xuelian Ge, Jian Weng, Xiao Han, and Feiyan Chen(✉)

Bio-X Laboratory, Department of Physics, Zhejiang University, Hangzhou 310027, China
chenfy@zju.edu.cn

Abstract. Influences of age and hemispheric asymmetry are important to white matter (WM) development, which are related with cognitive function development during childhood. However, the brain WM developmental trajectory of Chinese children remains to be less fully investigated and possibly follows different trends compared with Western children, since there have been behavior or cognitive evidence suggesting the intercultural developmental differences between them. In this study, we utilized the cross-sectional diffusion tensor imaging technique to measure the age-related WM changes and hemispheric asymmetry in 71 right-handed Chinese children ranging in age from 6 to 10 years. DTI indices of fractional anisotropy (FA), mean diffusivity (MD), as well as axial diffusivity (AD) and radial diffusivity (RD) were assessed in 12 major white matter tracts in both two hemispheres respectively. We found the developmental age correlated positively with FA value and negatively with MD value in most WM regions observed. Besides, hemispheric asymmetry was identified in most of WM tracts. By contrast, the age-related WM changes and hemispheric asymmetry found in Chinese children population may be more widespread than in Western children population reported by previous studies. Overall, the present study characterized the influence of age and hemispheric asymmetry on WM of Chinese children's brain during development.

Keywords: White matter maturation · Hemispheric asymmetry · Chinese children

1 Introduction

Previous evidence from structural MRI revealed that human brain undergoes continuous developmental changes from infancy to young adulthood. It generally believed that brain white matter (WM) grows considerably in the first few years of life, while volumetric studies have demonstrated that WM maturation in late childhood and even in adolescent and young adulthood, though steady and regional, is also significant and has drawn increasing scientific interests [13]. Diffusion tensor imaging (DTI) provides a non-invasive way to characterize white matter growth by measuring primary indices, e.g., the fractional anisotropy (FA) and mean diffusivity (MD), which can quantitatively

X. Ying (Ed.): HBAI 2022, CCIS 1692, pp. 195–207, 2023.
https://doi.org/10.1007/978-981-19-8222-4_17

describe the fiber bundle structure based on microscopic water motion [2]. Due to its high sensitivity and specificity on WM, DTI has become a valid tool of measurement on the anatomical structure as well as the important maturation information of WM.

To date, DTI has been applied to investigate the long-term WM maturation from infancy to adulthood [14], and showed convergent evidence to increase FA and decreased MD value in WM regions. Specifically, most of diffusion changes were observed to occur within the period of neonates and infants [14, 19]. Meanwhile, age-related influence on diffusion indices of the major white matter tracts, subcortical white matter, and deep gray matter lasts afterwards until adolescence and even young adulthood [18, 32]. However, compared with the neonatal of infant and adolescence WM growth on which previous studies focused, age-related changes are poorly documented among children population [32].

On the other hand, the structural hemispheric asymmetry was also a striking aspect of brain structural features and development of human brain. Asymmetries of brain WM emerges as early as the fetal stage [25]. Leftward asymmetries in older ages (adolescence and adults) in the left corticospinal tract and left superior longitudinal fasciculus (SLF) [9], etc., were well documented and may be experience/environment related. However, the asymmetries of WM tracts in children population, which may be considered to reflect more genetic-related information, remain to be less investigated [4, 32].

It has been demonstrated that age-related WM maturation [3, 24] and hemispheric asymmetry [16, 26] are correlated closely with neurocognitive functions in certain aspects. Language ability contributed to asymmetry of SLF [26]. Meanwhile, evidence from behavior researches have suggested that the cognitive development pattern of children was development on intercultural differences. For example, the development difference was indicated between Chinese and Western children in language [29]. Therefore, it is possible that age-related white matter maturation and hemispheric asymmetry may have different trends in Chinese compared to previous studies based on Western children population.

Based on our collected DTI data of Chinese children population, there are two hypotheses. First, the developmental trajectory of WM during the childhood period of 6–10 years could be characterized based on diffusion indices. The second hypothesis is that substantial hemispheric asymmetry in white matter tracts will be observed, for the reason of cognitive function development in Chinese children population.

2 Method

2.1 Participants

Seventy-one children attended the present cross-sectional study: mean age ± SD, 7.71 ± 0.80 years; age range, 5.95 – 9.75 years; 37 girls. All children were right-handed. They were all native Chinese speakers and excluded for any psychiatric or neuropsychiatric disorders.

2.2 Image Acquisition

The MRI data were acquired on a 1.5T MR scanner (Achieva, Philips) equipped with an eight-channnel Philips SENSE head coil. The DTI data were obtained with the single-shot echo planar imaging sequence (repetition time 1300 ms; echo time 77 ms; field-of-view 256 × 256 mm^2; acquisition matrix 128 × 128, 64 slices, slice thickness 2 mm, no gap). The sequence was performed with one non-diffusion-weighted image ($b = 0$ s/mm^2) in addition to 32 diffusion weighted images ($b = 1000$ s/mm^2) collected.

2.3 Image Analysis

Image analysis and tensor calculation were done by using FMRIB's diffusion toolbox (FSL, version 4; www.fmrib.ox.ac.uk/fsl, [31]). Eddy current distortion and motion artifacts in the DTI dataset were corrected by affine registration to the non-diffusion image. Then non-brain tissue and background noise were removed using the Brain Extraction Tool (BET v2). The eigenvalues ($\lambda1$, $\lambda2$, and $\lambda3$) were obtained by diagonalizing the tensor matrix. The diffusion indices fractional anisotropy (FA), mean diffusion (MD), axial diffusivity (AD) and radial diffusivity (RD) of each voxel were calculated using DTIfit within the FMRIB Diffusion Toolbox.

Preprocessed diffusion data were analyzed using TBSS (Tract-Based Spatial Statistics, [30]). After aligned to the FMRIB_58 template in standard MRI152 space using the FMRIB's Nonlinear Registration Tool, each subject's FA map was projected onto the standard FMRIB_58 skeleton and the skeleton with the threshold set at $FA > 0.20$.

Mean, axial and radial diffusivity skeletons were constructed with the skeleton-projection parameters estimated from FA skeleton procedure using the tbss_non_FA procedure provided in FSL.

2.4 Tractography Regions of Interest

Diffusion values (FA, MD, AD, and RD) were calculated for each participant's prior ROIs which were created based on probabilistic WM atlases from the JHU-ICBM-DTI-81 white matter labels atlas [12]. Six white matter tracts delineated as ROIs in each hemisphere (12 ROIs in all) were as follows: cingulum cingulate gyrus (CgC), cingulum hippocampus (CgH), corticospinal tracts (CST), superior longitudinal fasciculi (SLF), superior fronto-occipital fasciculi (SFOF), and uncinate fasciculus (UF) (shown in Fig. 1). Mean values for FA, MD, AD, and RD were obtained on the skeleton within each region.

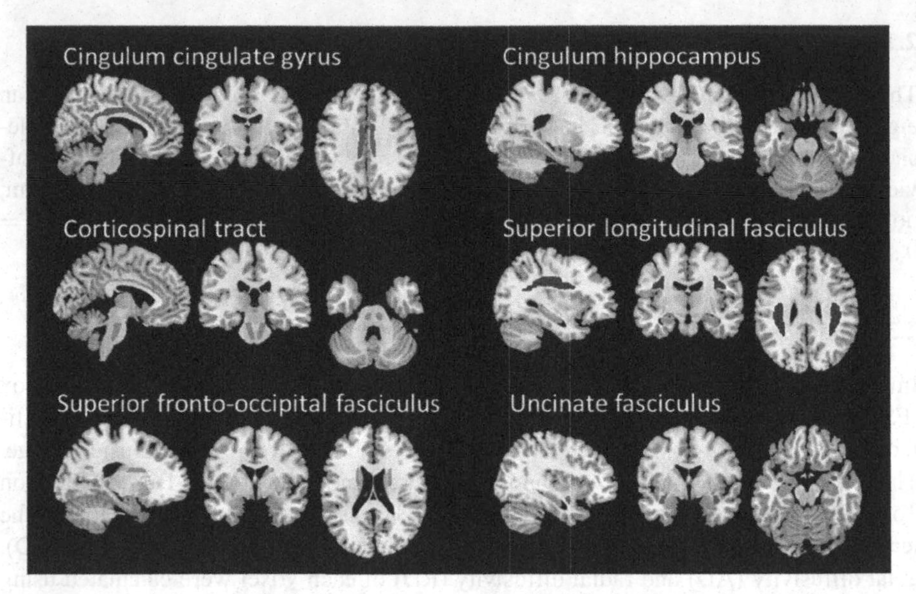

Fig. 1. Bilateral white matter tracts examined.

2.5 Statistical Analysis

Statistical analysis was performed using Statistical Package for the Social Sciences (SPSS25.0, http://www.spss.com). The relations between FA, MD, AD, RD values of each ROI and age were investigated using Pearson's correlation. Hemispheric asymmetry analysis was assessed by paired t-tests. Results were considered significant at $p < 0.05$.

3 Results

3.1 Age-Related White Matter Maturation

The DTI indices of all twelve white matter tracts followed similar trend of changes during the childhood period: FA gradually increased with age, MD and RD decreased with age, and AD stayed constantly. Specifically, the correlation between FA and age was significantly in 9 of 12 tracts (Fig. 2). On the other hand, a significant decrease in MD was found in 10 of 12 tracts (Fig. 3). As for the eigenvalues of diffusion tensor, the AD values stayed constantly, only bilateral CST decreased significantly with age (Fig. 4). While a significant negative correlation between RD values and age was noted in 9 of 12 tracts (Fig. 5). We also found no tracts of white matter in which FA decreased or MD increased with age.

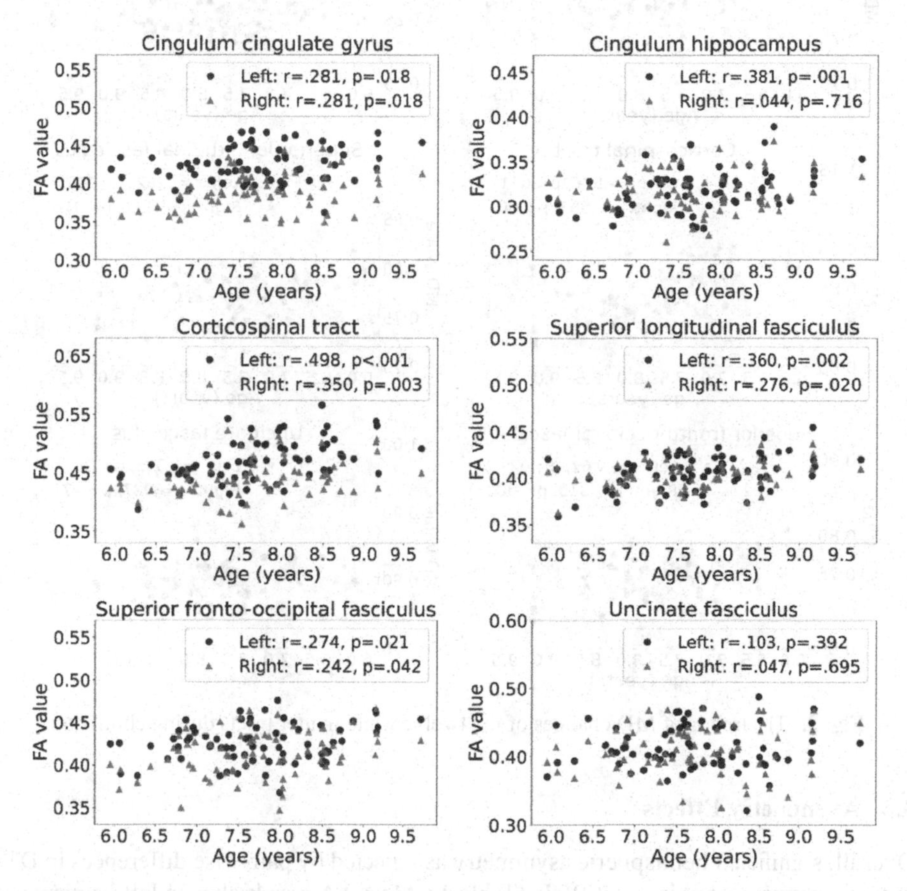

Fig. 2. The trends of FA changes of the twelve white matter tracts during childhood.

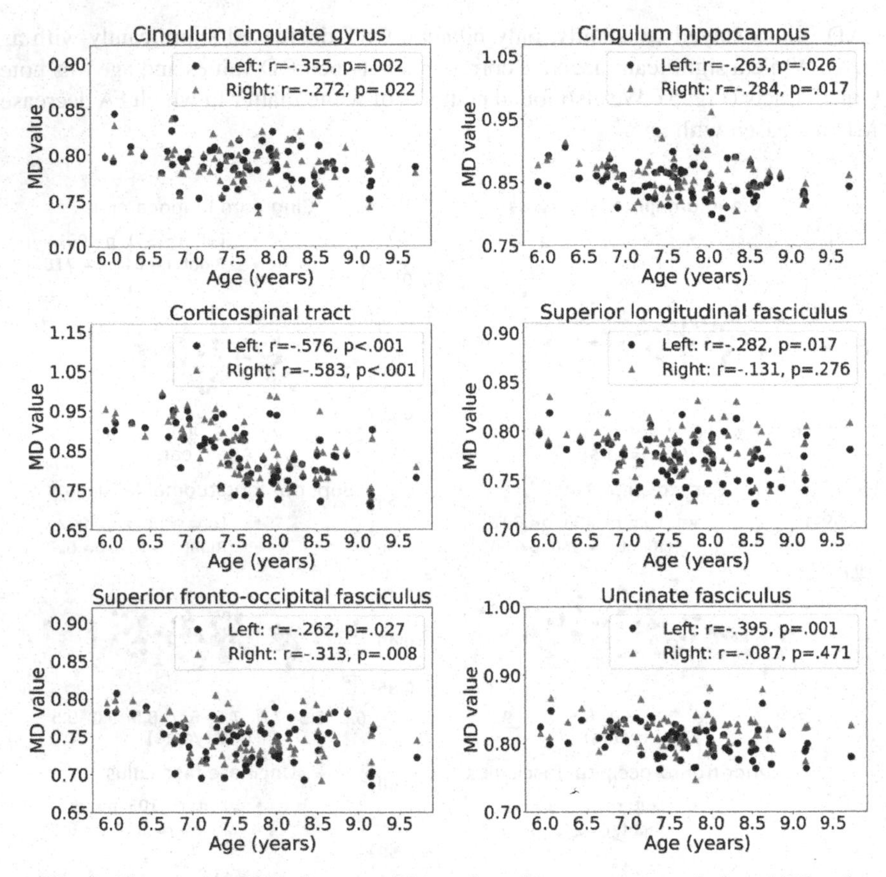

Fig. 3. The trends of MD changes of the twelve white matter tracts during childhood.

3.2 Asymmetry Effects

Overall, significant hemispheric asymmetry as reflected by pair-wise differences in DTI indices was presented in most ROIs (Table 1). Mean FA was higher in left hemisphere but not for CgH and UF. Similarly, we found significantly leftward asymmetry of higher

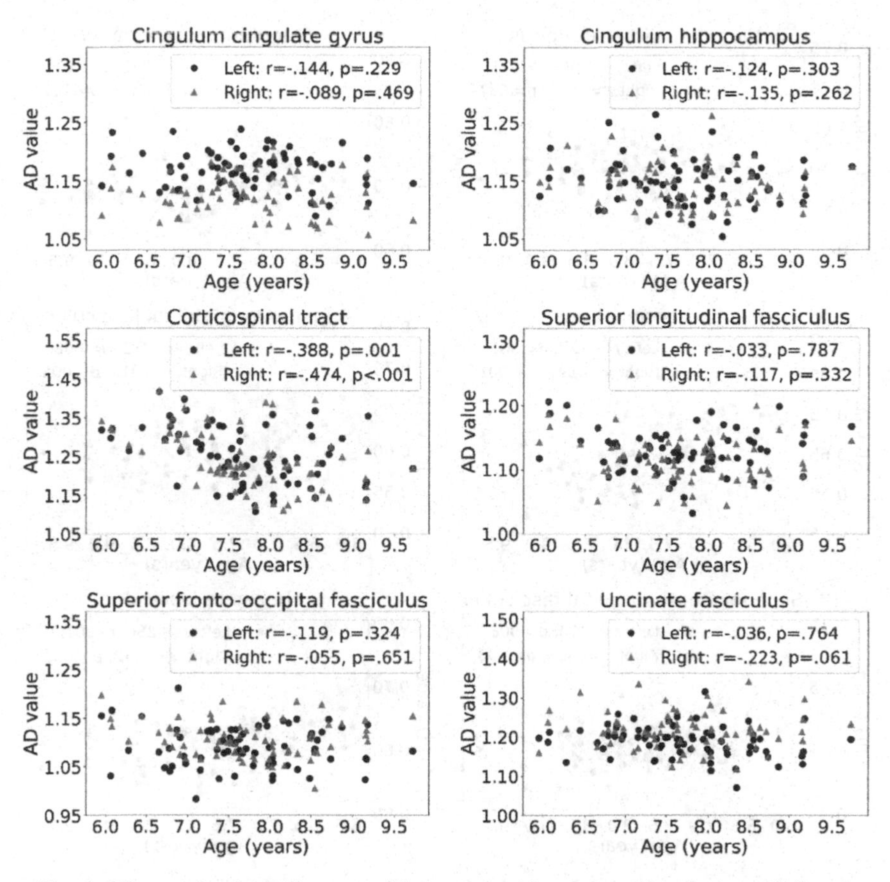

Fig. 4. The trends of AD changes of the twelve white matter tracts during childhood.

AD in CgC, CST and SLF, and AD was higher in right hemisphere for SFO and UF with the exception being in the CgH, which had no significant hemispheric asymmetry in AD. Mean MD was significantly greater in the right compare with left hemisphere, while this pattern was not found in CgC. The rightward asymmetry evaluation for RD values was significant in all tract pairs.

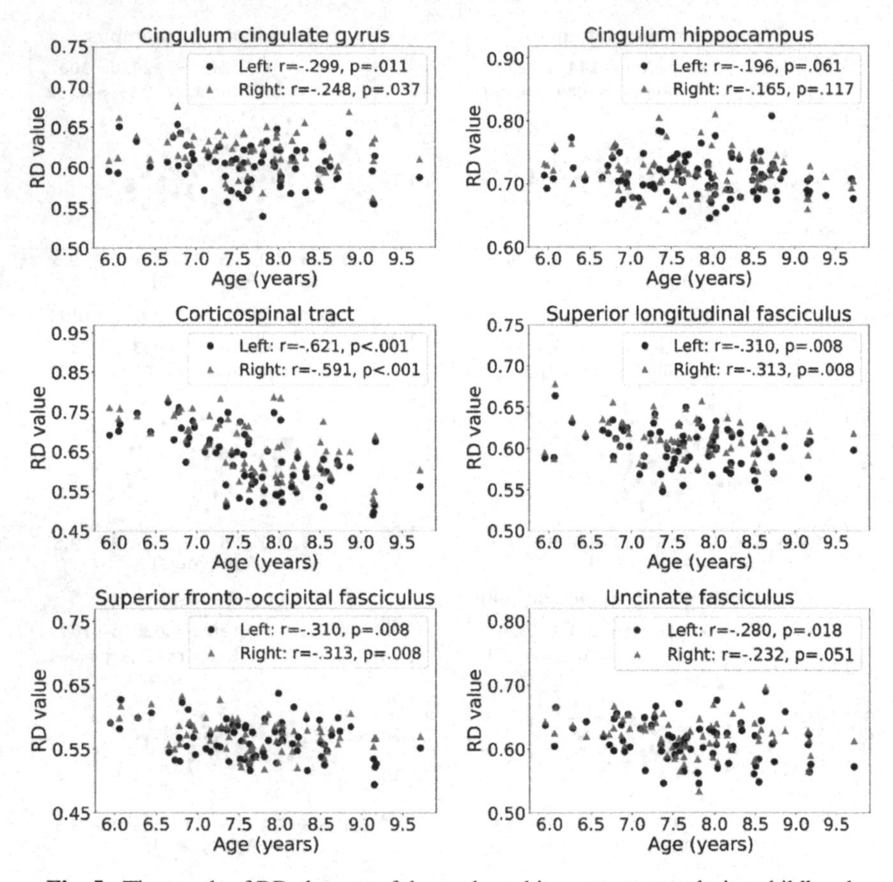

Fig. 5. The trends of RD changes of the twelve white matter tracts during childhood.

Table 1. Diffusion values (FA, MD, AD, RD) calculated for the twelve ROIs.

Tract name	Left	Right	Paired-p
Cingulum cingulate gyrus			
FA	0.43	0.39	< .01
MD	0.79	0.79	= .80
AD	1.17	1.13	< .01
RD	0.60	0.62	< .01
Cingulum hippocampus			

(*continued*)

Table 1. (*continued*)

Tract name	Left	Right	Paired-p
FA	0.32	0.32	= .89
MD	0.85	0.86	< .05
AD	1.15	1.14	= .58
RD	0.71	0.72	< .05
Corticospinal tract			
FA	0.47	0.44	< .01
MD	0.83	0.85	< .01
AD	1.25	1.24	<.05
RD	0.63	0.66	<.001
Superior longitudinal fasciculus			
FA	0.41	0.40	< .05
MD	0.77	0.78	< .01
AD	1.13	1.11	< .01
RD	0.60	0.61	< .01
Superior fronto-occipital fasciculus			
FA	0.43	0.41	< .01
MD	0.74	0.75	< .05
AD	1.09	1.11	< .01
RD	0.56	0.57	< .05
Uncinate fasciculus			
FA	0.42	0.41	= .21
MD	0.80	0.82	< .01
AD	1.19	1.22	< .01
RD	0.61	0.62	< .05

4 Discussion

In current study, we wanted to explore the age-related WM maturation and hemispheric asymmetry of normal Chinese children based on their DTI data. Three major findings were observed. Firstly, consistent with our hypothesis, we confirmed the widespread age-related DTI changes of increasing FA and decreasing MD in Chinese children, reflecting the process of brain WM maturation during childhood, with the observed changes of intensely decreasing RD combining with basically unchanged AD. In line with previous studies [15, 17], diffusion indices changes may suggest the underlying myelination process and water content loss occurred in WM. Secondly, compare with previous studies based on Western children population, the age-related WM maturation in Chinese children population was found to be more widespread. One of the possible reasons for this result is cross-cultural cognitive function development difference [28]. Thirdly, we found widespread significant hemispheric asymmetry in the term of diffusion indices in Chinese children's brain, which was in line with other asymmetry studies based on Western adolescence and adult population [4, 34]. In fact, asymmetry of white matter tracts was closely associated with complex cognitive function, such as the SLF [9, 20]. However, previous brain asymmetry studies did not report these regions. These differences between Chinese and Western children may have been related to the difference in cognitive function development.

4.1 Age-Related White Matter Maturation

Being consistent with previous investigation on age's effect on diffusion indices [10], we also found age-related FA increases, and MD and RA decreases in different WM regions of the brain in our Chinese children population. This trend of age's positive correlation with FA as well as negative correlation with MD during childhood were consistent with previous findings from infancy to adulthood [15]. However, in contrast to previous findings that only two or three tracts are age-related [4, 10], we found a broader tract showed age-related correlation with FA and MD. The reason for this distinction may be cause by the differences of cognitive function. Cultural psychologists have noted that Chinese children may show more mature of cognitive functions than Western children in the period of childhood, such as executive function [27] and language ability [29]. They pointed that Chinese parent expect children to learn language and to master other functions as young as they could, whereas Western parents do not expect such mastery during young childhood. This education strategy may lead to these cognitive functions developed in childhood of Chinese, earlier than Western population. Previous studies also have proven that age-related correlation with diffusion indices in white matter tracts thought linked to specific cognitive function development [22, 24]. Further correlations between white matter microstructure and cognitive functions were observed in the SLF associated with language ability [26], the cingulum associated with executive function [24]. Therefore, global age-related increases in FA and decreases in MD that possibly occurred more during Chinese children is not surprising.

Age-related increase in FA and decrease in MD might have been mainly caused by a decrease in RD [1, 4]. Among our children sample, we found widespread age-related decrease in RD, while the age effects on AD were only found significant in bilateral CST, which were similar to the diffusivity's changes across lifespan [15]. A decrease in RD combined with basically unchanged AD was known to be reflective of the myelination process [7]. And the myelination process is believed to lead to the increases of FA and decreases of MD during white matter maturation [4, 7], which is also in line with our current findings of FA and MD. Moreover, the present findings of decreasing RD in left SLF and left UF may also reflect the developmental pattern of glial cell bodies and membranes proliferate, which both lead to a decrease in brain water content [7]. In agreement with developmental studies, reduction of water content is an additional sign of brain maturation that could contribute to the changes in FA and MD [4, 7, 8].

Compared with other WM tracts, the bilateral UF showed non-significant developmental changes. According to previous studies, some white matter tracts which play a role in higher level cognitive functions may develop late and continued to mature in young adulthood [17, 28]. Take the UF for instance, it connects the temporal lobe with the frontal lobe and composes a critical part of ventral language pathway [23]. Further studies also revealed its potential role in auditory working memory [21] and semantic processing [6]. Since the development of this function generally established until adolescence, it explained the comparatively less significant age-related FA changes in UF during childhood in our study [17, 28]. Therefore, our finding may support the trend of late development of UF in Chinese children.

4.2 Asymmetry Effects

Previous studies documented that significant hemispheric asymmetry was presented in adults [16, 34], but also in newborn infants [9], while the children population remain to be less investigated. In this study, we found significant asymmetry of FA values in four tract pairs, in line with previous studies based on adolescence and adults [34]. All the asymmetries in FA values were leftward in our Chinese children population. Similar to age-related white matter maturation, studies have indicated a potential link between hemispheric asymmetry and cognitive functions [9, 20]. Thus, these currently suggested differences in hemispheric asymmetry may underlie the behavior or cognitive differences between Chinese and Western children.

Language is an extremely complex cognitive function of human and is lateralized in terms of both brain function and structure. Studies have reported consistently that the SLF plays an important role in language function [16, 26]. For example, comprehensive studies proposed that the SLF supports language function of speech processing and production, as well as reading and semantic processing [6, 11]. Moreover, these studies also demonstrated that language is predominantly processed in the left hemisphere. This is probably one of the remarkable reasons for human brain leftward asymmetry and may be correlated closely with our WM findings of asymmetry pattern in SLF. Compared to significant leftward FA asymmetry in our findings at the age range of 6 to 8 years, no asymmetry in SLF was reported during the infant and childhood periods of Western population [33]. Meanwhile, another study based on Chinese population [26], reported this asymmetry and concluded that it may considered to be associated with language development of Chinese children. In China, children generally undergo intense language practice during early school age (around 6 – 8 years), which may intensively activate language functions of various categories [5, 29]. This may contribute to a more widespread and significant leftward asymmetry observed an early age in Chinese children cohort in our study.

5 Conclusions

The present study has characterized the influence of age and hemispheric asymmetry on WM of Chinese children' brain. We found the developmental age correlated positively with FA value and negatively with MD value in most WM regions observed in Chinese children's brain. Taken together with the observed changes of intensively decreasing RD combing with basically unchanged AD, our findings may suggest that underlying myelination process and water content loss occurred in WM. Besides, leftward hemispheric asymmetry was identified in most of WM tracts. Compared with previous studies on Western children population, the age-related WM changes and hemispheric asymmetry found in Chinese children may be more widespread, which may be possibly related to the intercultural cognitive or behavior developmental differences between them.

Acknowledgements. We are grateful to the Chinese Abacus and Mental Arithmetic Association and Heilongjiang Abacus Association for their kind support, as well as to the children, parents, and teachers of Qiqihar for their participation in the study. This work was funded by the National Natural Science Foundation of China.

References

1. Huang, H., Zhang, J.Y., Wakana, S., et al.: White and gray matter development in human fetal, newborn and pediatric brains. Neuroimage **33**(1), 27–38 (2006)
2. Beaulieu, C.: The basis of anisotropic water diffusion in the nervous system - a technical review. NMR Biomed. **15**(7–8), 435–455 (2002)
3. Johnson, R.T., Yeatman, J.D., Wandell, B.A., et al.: Diffusion properties of major white matter tracts in young, typically developing children. Neuroimage **88**, 143–154 (2014)
4. Lehtola, S.J., et al.: Associations of age and sex with brain volumes and asymmetry in 2–5-week-old infants. Brain Struct. Funct. **224**(1), 501–513 (2018). https://doi.org/10.1007/s00429-018-1787-x
5. Lebel, C., Denoi, S.: The development of brain white matter structure. Neuroimage **182**, 207–218 (2018)
6. Snook, L., Paulson, L.A., Roy, D., Phillips, L., Beaulieu, C.: Diffusion tensor imaging of neurodevelopment in children and young adults. Neuroimage **26**(4), 1164–1173 (2005)
7. Provenzale, J.M., Liang, L., DeLong, L., White, L.E.: Diffusion tensor imaging assessment of brain white matter maturation during the first postnatal year. Am. J. Roentgenol. **189**(2), 476–486 (2007)
8. Dubois, J., Hertz-Pannier, L., Cachia, A., et al.: Structural asymmetries in the infant language and sensori-motor networks. Cereb. Cortex **19**(2), 414–423 (2009)
9. Bonekamp, D., Nagae, L.M., Degaonkar, M., et al.: Diffusion tensor imaging in children and adolescences: reproducibility, hemispheric, and age-related differences. Neuroimage **34**(2), 733–742 (2007)
10. Bells, S., Lefebvre, J., Longoni, G., et al.: White matter plasticity and maturation in human cognition. Gila **67**(11), 2020–2037 (2019)
11. Peters, B.D., et al.: Age-related differences in white matter tract microstructure are associated with cognitive performance from childhood to adulthood. Biol. Psychiat. **75**(3), 248–256 (2014)
12. Lebel, C., Beaulieu, C.: Lateralization of the arcuate fasciculus from childhood to adulthood and its relation to cognitive abilities in children. Hum. Brain Mapp. **30**(11), 3563–3573 (2009)
13. Qiu, D.Q., Tan, L.H., Siok, W.T., Zhou, K., Khong, P.L.: Lateralization of the arcuate fasciculus and its differential correlation with reading ability between young learners and experienced readers: a diffusion tensor tractography study in a Chinese cohort. Hum. Brain Mapp. **32**(12), 2054–2063 (2011)
14. Siok, W.T., Niu, Z.D., Jin, Z., Perfetti, C.A., Tan, L.H.: A structural–functional basis for dyslexia in the cortex of Chinese readers. Proc. Natl. Acad. Sci. **105**(14), 5561–5566 (2008)

15. Smith, S.M., et al.: Advances in functional and structural MR image analysis and implementation as FSL. Neuroimage **23**, S208–S209 (2004)
16. Smith, S.M., et al.: Tract-based spatial statistics: voxelwise analysis of multi-subject diffusion data. Neuroimage **31**(4), 1487–1505 (2006)
17. Hua, K., Zhang, J.Y., Wakana, S., et al.: Tract probability maps in stereotaxic spaces: analysis of white matter anatomy and tract-specific quantification. Neuroimage **39**(1), 336–347 (2008)
18. Lebel, C., Walker, L., Leemans, A., Phillips, L., Beaulieu, C.: Microstructural maturation of the human brain from childhood to adulthood. Neuroimage **40**(3), 1044–1055 (2008)
19. Lebel, C., Gee, M., Camicioli, R., Wieler, M., Martin, W., Beaulieu, C.: Diffusion tensor imaging of white matter tract evolution over the lifespan. Neuroimage **60**(1), 340–352 (2012)
20. Simmonds, D.J., Hallquist, M.N., Asato, M., Luna, B.: Developmental stages and sex differences of white matter and behavioral development through adolescence: a longitudinal diffusion tensor imaging (DTI) study. Neuroimage **92**, 356–368 (2014)
21. de Schotten, M.T., et al.: Atlasing location, asymmetry and inter-subject variability of white matter tracts in the human brain with MR diffusion tractography. Neuroimage **54**(1), 49–59 (2011)
22. Liu, Y., et al.: Structural asymmetries in motor and language networks in a population of healthy preterm neonates at term equivalent age: a diffusion tensor imaging and probabilistic tractography study. Neuroimage **51**(2), 783–788 (2010)
23. Giorgio, A., Watkins, K.E., Chadwick, M., et al.: Longitudinal changes in grey and white matter during adolescence. Neuroimage **49**(1), 94–103 (2010)
24. Sabbagh, M.A., Xu, F., Carlson, S.M., Moses, L.J., Lee, K.: The development of executive functioning and theory of mind a comparison of Chinese and US preschoolers. Psychol. Sci. **17**(1), 74–81 (2006)
25. Nagy, Z., Westerberg, H., Klingberg, T.: Maturation of white matter is associated with the development of cognitive functions during childhood. J. Cogn. Neurosci. **16**(7), 1227–1233 (2004)
26. Ashtari, M., Cervellione, K.L., Hasan, K.M., et al.: White matter development during late adolescence in healthy males: a cross-sectional diffusion tensor imaging study. Neuroimage **35**(2), 501–510 (2007)
27. Dubois, J., Dehaene-Lambertz, G., Kulikova, S., et al.: The early development of brain white matter: a review of imaging studies in fetuses, newborns and infants. Neuroscience **276**, 48–71 (2014)
28. Dubois, J., Dehaene-Lambertz, G., Perrin, M., et al.: Asynchrony of the early maturation of white matter bundles in healthy infants: quantitative landmarks revealed noninvasively by diffusion tensor imaging. Hum. Brain Mapp. **29**(1), 14–27 (2008)
29. Parker, G.J.M., Luzzi, S., Alexander, D.C., et al.: Lateralization of ventral and dorsal auditory-language pathways in the human brain. Neuroimage **24**(3), 656–666 (2005)
30. McDonald, C.R., et al.: Diffusion tensor imaging correlates of memory and language impairments in temporal lobe epilepsy. Neurology **71**(23), 1869–1876 (2008)
31. Dick, A.S., Tremblay, P.: Beyond the arcuate fasciculus: consensus and controversy in the connectional anatomy of language. Brain **135**(12), 3529–3550 (2012)
32. Glasser, M.F., Rilling, J.K.: DTI tractography of the human brain's language pathways. Cereb. Cortex **18**(11), 2471–2482 (2008)
33. Song, J.W., Mitchell, P.D., Kolasinski, J., Grant, P.E., Galaburda, A.M., Takahashi, E.: Asymmetry of white matter pathways in developing human brains. Cereb. Cortex **25**(9), 2883–2893 (2015)
34. Booth, J.R., Lu, D., Burman, D.D., et al.: Specialization of phonological and semantic processing in Chinese word reading. Brain Res. **1071**(1), 197–207 (2006)

A Digital Gaming Intervention Combing Multitasking and Alternating Attention for ADHD: A Preliminary Study

Jiaheng Wang[2,3], Mengyi Bao[6], Wenyu Li[5], Ji Wang[5], Kewen Jiang[6], Lin Yao[1,2,3(✉)], and Yueming Wang[3,4]

[1] Department of Neurobiology, Mental Health Center and Hangzhou Seventh People's Hospital, Zhejiang University School of Medicine, Hangzhou, China
lin.yao@zju.edu.cn
[2] MOE Frontier Science Center for Brain Science and Brain-machine Integration, Zhejiang University, Hangzhou, China
[3] The College of Computer Science, Zhejiang University, Hangzhou, China
[4] Qiushi Academy for Advanced Studies (QAAS), Zhejiang University, Hangzhou, China
[5] Shanghai Shuyao Information Technology Co., Ltd, Shanghai, China
[6] The Children's Hoapital Zhejiang University School of Medicine, Hangzhou, China

Abstract. Objective. Digital therapeutics are considered as promising alternatives for the treatment of attention-deficit hyperactivity disorder (ADHD). We developed a digital intervention combing multitasking and alternating attention, namely DTFAD001. DTFAD001 targets four dimensions of attention control and is designed in a video game-like format. This preliminary study aims to investigate feasibility and efficacy of DTFAD001 in children with ADHD. **Methods.** 20 eligible patients were randomized 1:1 to receive DTFAD001 or the control (a single task version of DTFAD001). Both groups administrated 4 weeks of intervention with 30 min per day, 5 days per week. The primary outcome was the mean change in TOVA API from pre-intervention to post-intervention. Feasibility was assessed by adverse events, patient compliance, and acceptability. **Results.** The patients' mean age was 8.27 years (SD = 0.96), with 15 males and 5 females. The experimental and control group improved TOVA API by 1.38 (SD = 2.25) and 1.31 (SD = 2.46), respectively. There was high compliance with intervention and no serious adverse events were reported. **Conclusions.** The study verified feasibility and effectiveness of DTFAD001. Further large-scale studies are needed to reveal statistically sound effects and investigate specific benefits from multitask-alternating.

Keywords: Digital intervention · ADHD · Multitasking

1 Introduction

Attention-deficit hyperactivity disorder (ADHD) is a prevalent, impairing condition that is frequently comorbid with other psychiatric disorders and creates

X. Ying (Ed.): HBAI 2022, CCIS 1692, pp. 208–219, 2023.
https://doi.org/10.1007/978-981-19-8222-4_18

a substantial burden for the individual, their family, and the community [15]. It has been estimated nearly 7.2% of children worldwide are diagnosed with ADHD symptoms [16]. The treatment options for pediatric patients with ADHD are often limited and inaccessible. On the one hand, medication-based interventions have shown short-term efficacy and might cause side effects which are the most concerns of caregivers. On the other hand, non-pharmacological interventions such as behavioral interventions demand lots of human and financial resources which put further pressures on the families of patients.

In the last two decades, computerized training programs targeting treatment of cognitive abilities have emerged and shown promising results [12]. Among them, neurofeedback training is widely investigated to treat children with ADHD based on the idea of neural plasticity [10,11]. Moreover, training of working memory associated with prefrontal functioning has received much attention for ameliorating the symptoms in ADHD [7]. More recently, video game-based training paradigms specialized in activation of cognitive functioning promise to be the alternatives for the treatment of ADHD [4,8]. Angura et al. first designed a racing game combined with multitasking for the training of cognitive control in older adults [1]. Their adaptive, interference-rich video game was capable of improving multitasking and other untrained cognitive control abilities in older adults after one month intensive training. The key mechanism behind the game is multitasking targeting the management of cognitive interference which occurs when two or more tasks compete for cognitive and attention resources. What's more, an adaptive staircase algorithm was leveraged to maintain both equivalent difficulty and engagement in the component tasks which was crucial for effective and engaging training. The above findings further motivated the high-level development of a serious game aiming at a novel digital therapeutic for ADHD patients, AKL-T01 (Akili Interactive Labs, Boston, MA, USA) [9]. The trials on both medication-free and medication-treated pediatric ADHD populations confirmed AKL-T01 can be used to improve ADHD symptoms while safer and more acceptable. Nevertheless, it is necessary to validate the efficacy of this novel digital intervention in a broader population worldwide and investigate how to adapt task components to drive maximal neural and cognitive benefits. Besides, it is unclear whether children with ADHD benefit more from multitasking compared with single task training.

Considering the above challenges, in this paper, we develop a novel digital intervention, DTFAD001 (Chuanqi Research and Development Center of Zhejiang University, Hangzhou, China). DTFAD001 shares the majorities of design principles with AKL-T01 as it incorporates multitasking and challenge adaptivity in a video game-based training program. To promote training of broader attention dimensions, we further combined multitasking with alternating attention by means of administrating two multitasking components alternately, denoted as multitask-alternating. Specifically, previous studies are mainly designed with concurrent multitasking in which divided and selective attention systems are required to process several tasks simultaneously, while omitting interleaved multitasking that demands alternating attention which is one of

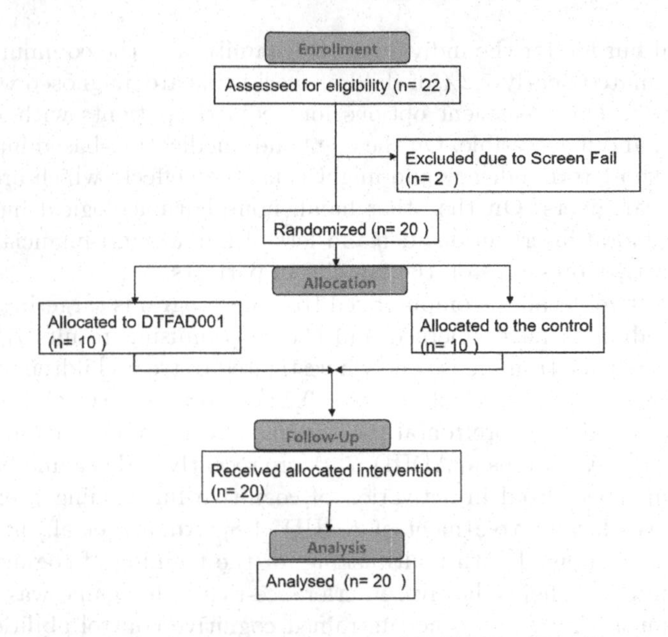

Fig. 1. The participant flowchart (CONSORT flowchart).

key aspects of attention control [5]. Hence, we introduce multitask-alternating in a game fashion to cover all aspects of attention control including selective attention, divided attention, alternating attention, and sustained attention.

The primary objective of this pilot study is to evaluate the feasibility and efficacy of DTFAD001 in children with ADHD. Besides, a control group treated with a single task version of DTFAD001 was examined to ablate contributions of multitasking compared with single task training.

2 Methods

This study was supported by a grant from Chuanqi Research and Development Center of Zhejiang University and was approved by the ethics review board of The Children's Hospital Zhejiang University School of Medicine. Written informed consent from parents and assent from children were obtained prior to study entry.

2.1 Trial Design

We conducted a single-center, randomized, double-blind, controlled study at the The Children's Hospital Zhejiang University School of Medicine in China from June 2021 to September 2021. Eligible patients were randomly allocated to either the experimental group or the active control group. The experimental group received intensive training using DTFAD001 for one month, while a single task version of the same game served as a mechanistic active control for the active control group. The current trial is a proof-of-concept study investigating in a small group of ADHD patients, exploring an emerging area of digital medicine.

2.2 Participants

We recruited pediatric patients aged 6 to 12 years who were diagnosed with ADHD based on the Diagnostic and Statistical Manual of Mental Disorders-Fourth Edition (DSM-IV). At pre-study screening, patients were required to undergo an additional continuous performance test (CPT), the Test of Variables of Attention (TOVA) [6]. Patients with TOVA Attention Performance Index (TOVA API) of −1.8 and below were included in the study, indicating cognitive deficits which were closely associated with ADHD symptoms. Children who recently experienced pharmacotherapy for ADHD were acceptable, provided they underwent a washout period for at least 7 days before baseline assessment to eliminate effects of medication. The key exclusion criteria were intellectual disability, co-existing psychiatric disorders, and other critical deficits that disrupted the normal course of training. Parents and patients were informed they were to receive one of two similar interventions and there was no evidence that one was superior to the other. Figure 1 provides details about the participant flow throughout the course of the study. We recruited 10 eligible patients for each of the intervention groups, and all of them accomplished a scheduled training with repeated outcome measures before and after the intervention.

2.3 Randomization and Blinding

The randomization sequence was generated by Chuanqi Research and Development Center of Zhejiang University using validated computerized pseudo-random number generator. Eligible patients were randomly assigned 1:1 to receive DTFAD001 or the control. Upon the approval of authorized ADHD specialists, parents were instructed to complete written informed consent with patient's assent. Then, unmasked personnel assigned the patients to the treatment groups in accordance with the pre-generated randomization sequence. Parents, patients, and specialists completing outcome measure assessments were masked to treatment allocation.

2.4 Interventions

Patients recruited in the study were provided with MatePad tablets (Huawei, China) for at-home training with either DTFAD001 or the control. At the beginning of the training, unmask research assistants instructed patients to complete the game tutorial and remind patients and parents to abide by the training protocol throughout the study.

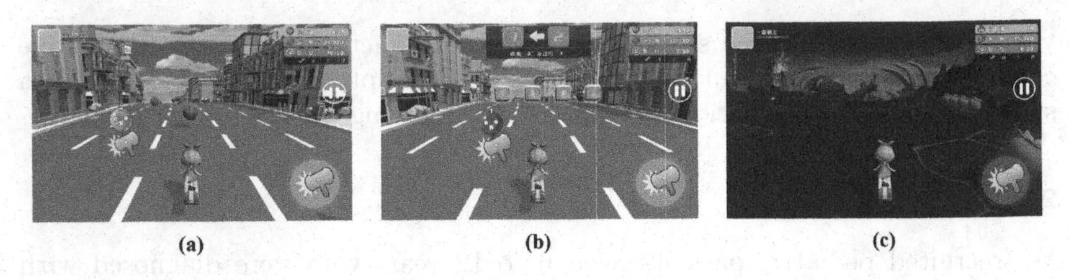

(a) (b) (c)

Fig. 2. Examples of DTFAD001 and the control. (a) and (b) represent two types of multitasking in DTFAD001, respectively. (c) represents a single task in the control.

The experimental group received DTFAD001 which was an investigational digital therapeutic targeting the treatment of attention-related cognitive deficits. DTFAD001 developed by Chuanqi Research and Development Center of Zhejiang University enjoys advantages of both rigorous scientific validation and high-level video game development. The key mechanisms behind DTFAD001 are multitask-alternating and challenge adaptivity. Multitask-alternating represents the alternate training of two multitasking components aiming at alternating attention which reflects the mental flexibility for fast concentration between different tasks. Two multitasking components are illustrated in Fig. 2. The first shown in Fig. 2a consists of a motor navigation task in which users continuously drive their avatars to collect items distributed on the scene as well as a perceptual discrimination task administrated in parallel. The latter is initialized with a stimulus item displayed frequently to serve as the interference wherein users ought to response to preferred items while ignore the other distractors. The second component replaces the collection task with the academic oriented task in which pediatric patients with ADHD often struggle in high quality performance due to impaired cognitive functioning. This is more associated with academic activities encountered by children and we hypothesize it will help children to generalize learned skills to real life. An illustrated example is given in Fig. 2b. A question is displayed near the top of the game scenario and users adjust the position of the avatar to meet the correct answer shown on the road. DTFAD001 administrates these two components alternately, covering the four dimensions of attention control and attempting to benefit related cognitive control abilities. Moreover, DTFAD001 is a closed-loop video game training system allowing the dynamic interactivity between the user and the game environment. Concretely, the speed of the avatar and stimulus items is adaptively updated on the basis of the real-time performance by means of a staircase algorithm proposed in [1]. This enables the appropriate difficulty to be specifically tailored to each individual and adjusted throughout the training progression. Furthermore, challenge adaptivity applies continuous pressure on the neural system activated by multitask-alternating, thus harnessing inherent neuro-plasticity processes and driving the desired neural changes [12]. Other feedback including earning rewards and unlocking new environments is presented to engage users.

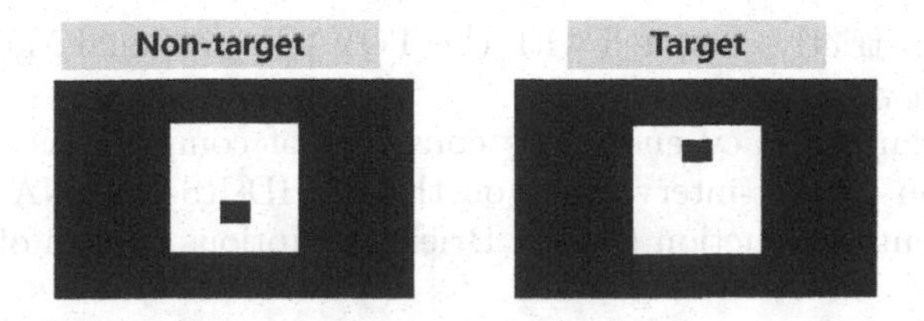

Fig. 3. The illustration of TOVA in two conditions.

The active control group underwent a single task version of DTFAD001 targeting to examine the effects of multitask-alternating compared with single task training. In contrast to multitasking components displayed in DTFAD001, the control only employs a collection task without any interference nor task-alternating, except a personalized training difficulty which is the same in DTFAD001. An epitome of the control game is shown in Fig. 2c. Patients and parents in the control group were not aware of receiving a simplified version of DTFAD001 and discouraged from discussing their randomized intervention with anyone other than an unmasked research assistant.

A 30 min per day, 5 days per week, 4-week intervention was administrated for both groups. In each day's intervention, patients completed 5 training sessions, approximately 6 min per session. Research assistants would track progress of intervention for each patient and remind those who broke off intervention for more than 48 h. In the event of any questions regarding the training procedures, patients and parents were able to contact the research assistants through phone and WeChat. The assessments were conducted before the intervention and after the completion of the training schedule to investigate efficacy on attention functioning, ADHD symptoms, and functional impairment.

2.5 Outcomes

The primary outcome measure was the mean changes in the TOVA API from pre-intervention to post-intervention in both groups. The TOVA is a validated, computerized, continuous performance test that objectively measures attention and impulse control processes in four areas: response time variability, response time, impulse control (commission errors), and inattention (omission errors). TOVA has been cleared by the US Food and Drug Administration (FDA) to facilitate assessment of attention deficits and commonly used in conjunction with other clinical tools or diagnostic tests in neuro-psychological or psychological evaluations. The procedure of the TOVA test is relatively simple. Subjects are instructed to respond to target stimuli flashing 0.2 s near the top of the screen as fast as possible, while not to react to nontarget stimuli appearing near the bottom of the screen. The illustration of two conditions is shown in Fig. 3. The TOVA API is a composite score of the sum of three scores: reaction time (RT) mean in Half-1 (highly infrequent targets), RT variability total in both halves, and d-prime in Half-2 (highly frequent targets). We developed the TOVA test program which strictly complied with the TOVA Professional Manual using the PsychoPy3 software [14].

The secondary efficacy endpoints consisted of comparisons of scores from pre-intervention to post-intervention on the ADHD-RS-IV, SNAP-IV AD/HD, Rating of Executive Function (REF). Brief descriptions of each of the measures are as follows:

- The ADHD-RS is an 18-item rating scale used to rate the frequency of ADHD symptoms based on the DSM-IV [13]. The score is a composite of the inattentive and hyperactive-impulsive symptoms: the higher the score, the more severe the symptoms.
- The SNAP is an 18-item rating scale used to evaluate the severity of inattentive and hyperactive-impulsive symptoms based on the DSM-IV [3]. There are two symptoms measured in SNAP that are the inattentive disorder (AD) and hyperactive-impulsive disorder (HD). The higher score corresponds to the worse condition.
- The REF is a parent-rated scale aiming at measures of impairment of executive function for children with ADHD. A higher score indicates a more serious functional impairment.

The feasibility of DTFAD001 is of the main concern in the study. Patients and caregivers were suggested to report any adverse events during the training period to their corresponding research assistants. Besides, research assistants would inquire about safety measures at the post-intervention visit. Details about use, performance, and compliance with intervention were automatically recorded by the study devices and uploaded to central servers when entering the game.

2.6 Data Analysis

The mean changes of TOVA API scores from pre-intervention to post-intervention was calculated on both groups. In addition, we performed the same analyses on the secondary efficacy endpoints, namely ADHD-RS, SNAP AD, SNAP HD, and REF. We also conducted between-group comparisons of changes on all measures. Statistical evaluation was not leveraged since insufficient number of samples investigated in this preliminary study. Specifically, statistical power analyses determined that a sample size of 44 patients per intervention group would be sufficient to detect an effect size of 0·5 with 90% or more power on a two-tailed, paired t-test. We intend to make analyses statistically significant in a large-scale study. All participants were included in the data analysis owing to high compliance with the training protocol.

To access feasibility, We obtained the number and proportion of participants reporting treatment-emergent adverse events during the intervention period. We also obtained the type, (worst) severity and frequency of treatment-emergent adverse effects. We report counts and proportions to assess safety of DTFAD001. What's more, the compliance with intervention is analyzed by calculating the proportion of required intervention sessions completed for each patient.

3 Results

3.1 Study Participation

A total of 20 participants were recruited in the study, including 15 males and 5 females with the mean age of 8.27 (SD = 0.96) and 7.8 (SD = 1.1), respectively. The trial commenced in July 2021, and ended in September 2021 as the last patient finished his intervention. Table 1 offers a summary of participants' demographic and clinical characteristics at the baseline. ALL participants accomplished their randomized intervention and got two assessments in day 0 (baseline) and day 30 (post-intervention). The post-intervention inquiries demonstrated the enjoyment of the training game.

Table 1. Baseline characteristics. Data are in the form of mean(SD).

	DTFAD001 (n = 10)	Control (n = 10)
Age	8.4 (1.0)	7.9 (0.9)
Male	8	7
Female	2	3
TOVA API	−4.0 (2.4)	−4.0 (1.3)
ADHD-RS	40.0 (9.8)	45.5 (7.0)
SNAP AD	1.4 (0.4)	1.9 (0.5)
SNAP HD	1.0 (0.3)	1.5 (0.7)
REF	0.9 (0.4)	1.0 (0.1)

3.2 Primary Outcomes

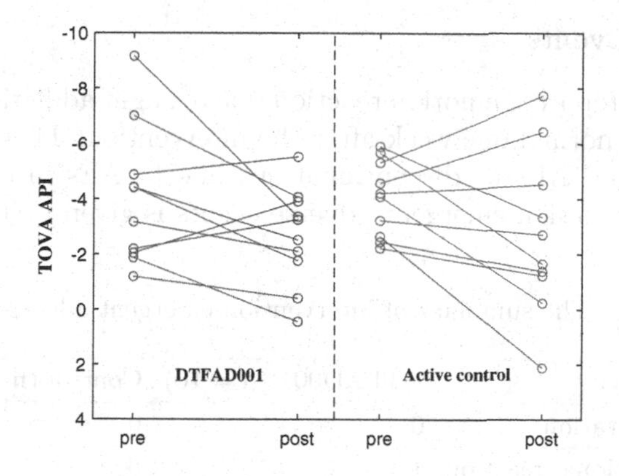

Fig. 4. TOVA API scores from pre-intervention to post-intervention in both groups.

Figure 4 illustrates the changes of TOVA API in two stages for both groups. The primary outcomes demonstrated the improvement from pre-intervention to post-intervention in both intervention groups. The mean change from the baseline on TOVA API for the experimental group was 1.38 (SD = 2.25) which was close to that for the control group (1.31, SD = 2.46).

3.3 Secondary Outcomes

Table 2 provides the assessment of changes referring to the mean differences of pre-post scores for both groups in all secondary endpoints: ADHD-RS, SNAP AD, SNAP HD, and REF. With regard to the experimental group, there were consistent improvements in scores. The similar reduction of scores can be observed in the control group. Patients randomly treated with DTFAD001 or the control both improved the ADHD-RS scores by 3.4 and 2.3 points, respectively. Moreover, SNAP AD and SNAP HD found changes of −0.21 and −0.13 in DTFAD001 as well as −0.22 and −0.28 in the control. There was also a slight improvement in executive functioning scores of 0.02 and 0.04 in the experimental group and the control group, respectively.

Table 2. Changes of secondary outcomes from pre-intervention to post-intervention. Data are in the form of mean(SD).

Outcome	DTFAD001 (n = 10)	Control (n = 10)
ADHD-RS	−3.40 (8.45)	−2.30 (10.30)
SNAP AD	−0.21 (0.42)	−0.22 (0.54)
SNAP HD	−0.13 (0.33)	−0.28 (0.66)
REF	−0.02 (0.33)	−0.04 (0.31)

3.4 Adverse Events

Only one patient reported emotion reaction due to slight addiction to DTFAD001 and recovered to normal one week after the intervention. There were no serious intervention-related AEs or discontinuations due to AEs in either group. The summary of intervention-emergent adverse events is given in Table 3.

Table 3. The summary of intervention-emergent adverse events.

AE type	DTFAD001 (n = 10)	Control (n = 10)
Frustration	0	0
Emotional reaction	1	0
Headache	0	0
Aggression	0	0

4 Discussion

In this preliminary study targeting to investigate the feasibility and efficacy of a digital intervention, the primary and secondary outcome measures provided preliminary support that DTFAD001 may be effective for improving attention functioning, ADHD symptoms, and executive functioning in children with ADHD. Apart from subjective measures which are the main references for the diagnosis of ADHD, TOVA serves as an objective measure of attention functioning. The comprehensive assessment of the effects produced by video game training verified the effectiveness of DTFAD001. The results of improved attention quantified by TOVA API were in agreement with previous studies which took similar forms of intervention [9]. Further, a slight improvement in executive functioning suggested a generalization effect to outcome measures that differed from target domains of intervention itself.

Just as important, the feasibility of DTFAD001 was convinced as no serious adverse events reported during the intervention period. Besides, there was high compliance with the training protocol for all patients. Compared with pharmacotherapy, DTFAD001 is safer and more acceptable while effective to treat ADHD symptoms. Moreover, DTFAD001 is more accessible to common families who have trouble in receiving behavioral or non-pharmacological interventions. As a novel digital therapeutic, DTFAD001 can be reasonably incorporated into a daily routine, utilized consistently by children with ADHD, and most critically implemented without the need for a physician, therapist, or clinician oversight.

Albeit a promising outcome of DTFAD001, the control showed similar effects on both primary and secondary efficacy measures. There are several factors to explain these findings. First, the only difference between DTFAD001 and the control is the number of tasks performed simultaneously, that is, multitask versus single task. While DTFAD001 targets border dimensions of attention control, they both involve sustained attention which is the core deficit of ADHD and share similar game mechanism. Thus, subjective measures might fail to capture minor changes benefited from multitask-alternating. More objective measures are needed to quantify fine-grained differences in cognitive functioning. Second, expectations of efficacy have been shown to moderate intervention effects in general, and also for digital interventions. In this pilot study, parents and patients were blinded to intervention groups and believed that they received a novel digital intervention for ADHD; hence, the close improvements could in part be due to expectations of efficacy arising in both groups. Finally, as reported in [2], multitasking training maintained an enhanced neural signature (midline frontal theta power) of cognitive control in a 6-year post intervention while the control didn't show that long-term effect. To that effect, long-term efficacy and durability are warranted to be examined in both groups. Current study only administrated a single intervention of one month duration, thereby the effects of a longer training schedule or additional booster sessions remain to be investigated.

In addition to the above considerations, several limitations are important to note. Firstly, we conducted a preliminary study to investigate feasibility and effectiveness of a digital intervention in a consumer-grade video game-like format. The number of patients investigated in the study is relatively insufficient to draw statistically sound outcomes. Large-scale randomized controlled trials based on the power statistical analysis is warranted to investigate the full clinical meaningfulness of current findings. Secondly, the strict inclusion and exclusion criteria omitted a broader population of patients with ADHD who might benefit from our digital therapeutic. Thirdly, it would be insightful to explain findings from the perspective of neural mechanistic. Neurophysiological signals such as EEG and FMRI can be utilized for mechanistic explanation underlying intervention effects. Moreover, it is of particular interest to combine video game and neurofeedback training.

5 Conclusion

DTFAD001 has demonstrated feasibility and effectiveness on the treatment of children with ADHD. This novel digital intervention is also safe, engaging and well-tolerated. Future generations of DTFAD001 promise to be sophisticated, multi-modal, targeted, and personalized.

Acknowledgements. We thank all volunteers for their participation in the study. This work was partly supported by grants from the National Key R&D Program of China (2018YFA0701400), Key R&D Program of Zhejiang (no. 2022C03011), the Chuanqi Research and Development Center of Zhejiang University, the Starry Night Science Fund of Zhejiang University Shanghai Institute for Advanced Study (SN-ZJU-SIAS-002), the Fundamental Research Funds for the Central Universities, Project for Hangzhou Medical Disciplines of Excellence, Key Project for Hangzhou Medical Disciplines, Research Project of State Key Laboratory of Mechanical System and Vibration MSV202115.

References

1. Anguera, J.A., Boccanfuso, J., Rintoul, J., Al-Hashimi, O., Faraji, F., Janowich, J., Kong, E., Larraburo, Y., Rolle, C.E., Johnston, E., Gazzaley, A.: Video game training enhances cognitive control in older adults. Nature **501**, 97–101 (2013)
2. Anguera, J.A., et al.: Long-term maintenance of multitasking abilities following video game training in older adults. Neurobiol. Aging **103**, 22–30 (2021)
3. Bussing, R., et al.: Parent and teacher snap-iv ratings of attention deficit hyperactivity disorder symptoms. Assessment **15**, 317–328 (2008)
4. Davis, N.O., Bower, J.D., Kollins, S.H.: Proof-of-concept study of an at-home, engaging, digital intervention for pediatric ADHD. PLoS ONE **13**, e0189749 (2018)
5. Douglas, H.E., Raban, M.Z., Walter, S.R., Westbrook, J.I.: Improving our understanding of multi-tasking in healthcare: Drawing together the cognitive psychology and healthcare literature. Appl. Ergon. **59**(Pt A), 45–55 (2017)
6. Forbes, G.B.: Clinical utility of the test of variables of attention (tova) in the diagnosis of attention-deficit/hyperactivity disorder. J. Clin. Psychol. **54**(4), 461–76 (1998)

7. Klingberg, T., Forssberg, H., Westerberg, H.: Training of working memory in children with ADHD. J. Clin. Exp. Neuropsychol. **24**, 781–791 (2002)
8. Kollins, S.H., Childress, A.C., Heusser, A.C., Lutz, J.: Effectiveness of a digital therapeutic as adjunct to treatment with medication in pediatric adhd. NPJ Digital Medicine 4 (2021)
9. Kollins, S.H., DeLoss, D.J., Cañadas, E., Lutz, J., Findling, R.L., Keefe, R.S.E., Epstein, J.N., Cutler, A.J., Faraone, S.V.: A novel digital intervention for actively reducing severity of paediatric ADHD (stars-ADHD): a randomised controlled trial. The Lancet. Digital health **2**(4), e168–e178 (2020)
10. Lim, C.G., et al.: A brain-computer interface based attention training program for treating attention deficit hyperactivity disorder. PLoS ONE **7**, e46692 (2012)
11. Lim, C.G., et al.: A randomized controlled trial of a brain-computer interface based attention training program for ADHD. PLoS ONE **14**, e0216225 (2019)
12. Mishra, J., Anguera, J.A., Gazzaley, A.: Video games for neuro-cognitive optimization. Neuron **90**, 214–218 (2016)
13. Pappas, D.N.: ADHD rating scale-iv: checklists, norms, and clinical interpretation. J. Psychoeduc. Assess. **24**, 172–178 (2006)
14. Peirce, J.W.: Psychopy-psychophysics software in python. J. Neurosci. Methods **162**, 8–13 (2007)
15. Posner, J., Polanczyk, G.V., Sonuga-Barke, E.J.S.: Attention-deficit hyperactivity disorder. Lancet **395**, 450–462 (2020)
16. Thomas, R., Sanders, S., Doust, J.A., Beller, E., Glasziou, P.P.: Prevalence of attention-deficit/hyperactivity disorder: a systematic review and meta-analysis. Pediatrics **135**, e1001–e994 (2015)

A BCI Speller with 120 Commands Encoded by Hybrid P300 and SSVEP Features

Xiaolin Xiao[1,2], Shengfu Wen[1], Jin Han[2], Man Yang[1], Erwei Yin[3,4], Minpeng Xu[1,2](✉), and Dong Ming[1,2]

[1] Academy of Medical Engineering and Translational Medicine, Tianjin University, Tianjin, China
minpeng.xu@tju.edu.cn

[2] Department of Biomedical Engineering, College of Precision Instruments and Optoelectronics Engineering, Tianjin University, Tianjin, China

[3] Defense Innovation Institute, Academy of Military Sciences (AMS), Beijing, China

[4] Tianjin Artificial Intelligence Innovation Center (TAIIC), Tianjin, China

Abstract. Implementing higher speed and larger command sets for brain-computer interfaces (BCIs) has always been the pursuit of researchers, which is helpful to realize the technological applications. The hybrid BCIs jointly induce different electroencephalogram (EEG) signals and could improve system performance effectively. This study designed an online BCI with 120 commands and high-speed by hybrid P300 and steady-state visual evoked potential (SSVEP) features. A time-frequency-phase encoding strategy was used to encode 120 commands in a short time, this strategy used time-locked P300s and frequency and phase-locked SSVEPs with a wide frequency band. The step-wise linear discriminant analysis (SWLDA) and ensemble task-related component analysis (eTRCA) were severally used to decode P300s and SSVEPs. As a result, online average spelling ac-curacy across six subjects was 83.89%. Average and highest information transfer rate (ITR) for this system was 151.53 bits/min and 175.09 bits/min, respectively. Meanwhile, the shortest time for out-putting one command was only 1.45 s. These results demonstrate the feasibility and effectiveness of this high-speed BCI with 120 commands, furthermore, this study used a wider frequency band of SSVEPs to encode 120 commands, which is helpful to extend larger command sets and achieve higher system performance.

Keywords: Brain-computer interface (BCI) · Hybrid BCI · P300 · Steady-state visual evoked potential (SSVEP) · High-speed · Large command sets

1 Introduction

Brain-computer interfaces (BCIs) can directly transform the activities of the central nervous system into artificial output, and realize direct interaction between the nervous system and the external environment [1–3]. In recent years, non-invasive electroencephalogram (EEG) has been widely used in BCIs due to its advantages of high temporal resolution and relatively low experimental cost. The commonly used EEG features are

X. Ying (Ed.): HBAI 2022, CCIS 1692, pp. 220–228, 2023.
https://doi.org/10.1007/978-981-19-8222-4_19

P300s [4], steady-state visual evoked potentials (SSVEPs) [5], sensorimotor rhythms (SMRs) [6], and their combinations [7].

The main criterion of BCI performance is information transfer rate (ITR), which is associated with the number of system commands, target recognition accuracy, and consumption time for each output command [8]. In order to build a BCI with high ITR, previous studies mainly paid attention to improving the recognition accuracy and reducing the consumption time instead of increasing the system commands. In 1988, Farwell and Donchin firstly designed a P300-speller with 30 commands that flashed according to the rows and columns encoding paradigm [9]. Since then, the number of commands in the BCI system has not increased obviously for a long time. It is worth noting that Jin et al. developed an adaptive P300-speller with a 12 × 7 command matrix in 2011, which extended the commands to 84 [10]. In 2018, Liu et al. proposed a new BCI paradigm based on code-modulated visual evoked potentials (c-VEPs) and achieved 64 commands [11]. Although extending the command set, these systems always prolonged the consuming time for outputting one command [12]. It is well known that there is a mutually restrictive relationship between the number of commands and target recognition accuracy and consuming time, hence it is still difficult to implement high-speed BCI systems with a large command set. In 2020, an ingenious coding method was proposed by Xu et al. [13], which combined time division multiple access (TMDA) in P300-speller with frequency division multiple access (FDMA) in SSVEP-BCI. This encoding method could provide more information and exhibited unique advantages [14, 15], which was used to propose a hybrid BCI with 108 commands. However, this BCI with 108 commands used a narrow SSVEP frequency band of 12.4 Hz–14.6 Hz, and ignored other available frequency bands for steady-state visual stimulation (SSVS) that had been used in high-speed SSVEP-BCIs [16]. Hence, exploring a wider frequency band of SSVEPs for a hybrid BCI could provide a basis for command set extension and system performance improvement.

Therefore, we explored the hybrid BCI based on P300 and SSVEP features with a wider SSVEP frequency band of 9 Hz–17.7 Hz, and realized an online high-speed BCI system with a larger command set. Specifically, we used a time-frequency-phase encoding strategy to encode 120 commands. These commands were arrayed by 30 parallel P300 sub-spellers, which were frequency-specific and contained a 2 × 2 command matrix. The step-wise linear discriminant analysis (SWLDA) and ensemble task-related com-ponent analysis (eTRCA) were used to decode hybrid P300 and SSVEP features. This online system could output one command in a short-time, and realized a hybrid BCI with both high-speed and large command sets.

2 Methods and Experiments

2.1 Subjects

Six healthy volunteers (aged 21 to 25 years, 2 females) participated in the study. All subjects had normal vision or were corrected to normal vision. Four of them had previously participated in the SSVEP-BCI or P300-BCI experiments, the remaining two subjects had no relevant experience. The Institutional Review Board of Tianjin University approved the experimental procedures in this paper, and each subject signed a written informed consent.

2.2 Hybrid Paradigm Design and Implementation

As shown in Fig. 1(a), 120 commands were presented on a 27-inch LCD monitor with a resolution of 1920 × 1080 pixels, and the refresh rate was 120 Hz. Each command corresponded to a 105 × 63 pixels stimulus square, and 120 commands were divided into thirty 2 × 2 matrices. Each command matrix was an independent P300 sub-speller. As shown in Fig. 1(c), the internal commands in each sub-speller were randomly flashed and traversed by the time encoding method. As shown in Fig. 1(b), the frequency-phase encoding method [16] was used to encode 30 P300 sub-spellers, with the frequency starting from 9 Hz to 17.7 Hz (the vertical interval was 0.3 Hz) and the phase started from 0 π (the horizontal interval was 0.35 π).

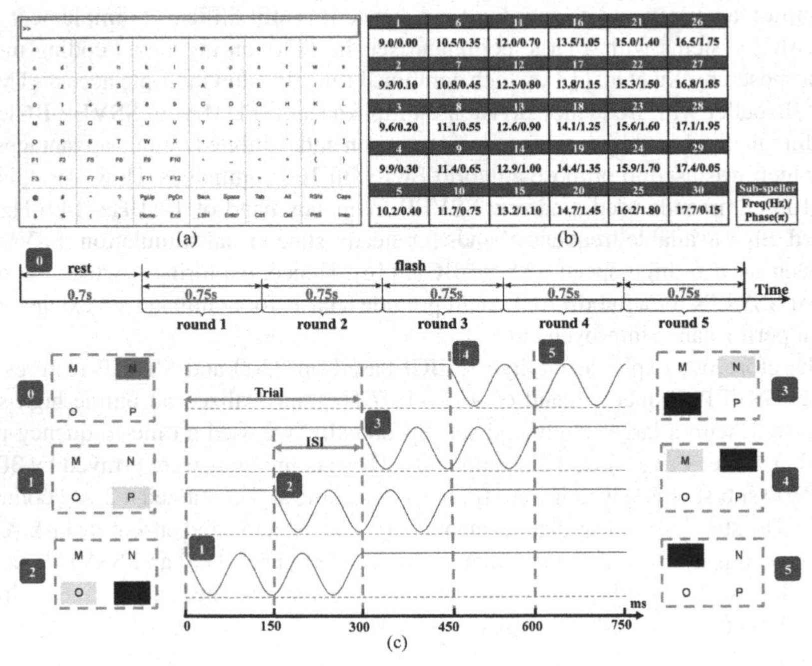

Fig. 1. Schematic diagram of the hybrid P300-SSVEP BCI paradigm. (a) 120 commands that were divided into 30 sub-spellers. (b) Frequency-phase encoding of one P300 sub-speller. (c) Stimulation process for one command spelling. The order of command flickering in the matrix was P- > O- > N- > M. The blue rounded rectangle containing a number was used to mark the time when the command stimulus appears. Here command 'N' was the target, while the other three commands were non-targets.

A complete stimulus process contained cueing stage and flickering stage. The cueing stage marked the target command with a red square, and the time lasted 0.7 s. In the flickering stage, all P300 sub-spellers flashed at the same time, the flickering time of each command was 0.3 s (recorded as a trial). The inter-stimulus interval (ISI) was – 0.15 s, hence the complete stimulus process of one command only used 0.75 s (recorded as a round). In the offline BCI experiment, all commands flashed continuously for five rounds, and the number of rounds in the online experiment was adjusted according to the subjects' offline system performance. The experimental program was developed with Psychophysics Toolbox Version 3 in the MATLAB environment, and each command was presented through the method of sampling sinusoidal stimuli.

2.3 BCI Experiment

The offline experiments contained 12 groups, each group cued 30 commands, and 4 groups of experiments would traverse all 120 commands. During the experiment, the subjects sat 60 cm in front of the monitor and were asked to seek and fixate on the target stimuli during the flickering stage. Most artifacts can be avoided to ensure the quality of the recorded signal in this way. Compared with the offline system, the online BCI system added the process of real-time signal analysis and recognition result feedback. Two computers in the system were used for visual stimulation (Windows 10, Intel Core i7–7700, RAM 16 GB) and data processing (Windows 7, Intel Core i7–7700, RAM 16 GB), respectively. In the online experiment, each subject was required to spell 30 commands, and the results would be fed back to the output box above the stimulation interface that shown in Fig. 1(a) in real-time.

2.4 EEG Recording and Processing

The EEG signals were collected by the Neuroscan SynAmps2 system, and the electrodes were distributed according to the international 10/20 system. The reference electrode was placed on the top of the head and the ground electrode was placed on the forehead. The system exhibited higher performance in this reference mode. To analyze P300 features, we selected FCz, Cz, Pz, PO7, PO8, and Oz around the parietal and occipital lobes [13]. Pz, PO3, PO4, PO5, PO6, POz, O1, O2, and Oz were selected for SSVEPs [17].

All recorded data were down-sampled to 250 Hz in pre-processing. For P300s, the signal in each channel was filtered to 1–15 Hz with a 3rd-order Butterworth band-pass filter. This passband filter effectively extracted P300 features while maintaining classification performance. The SSVEPs were analyzed by the filter bank method [18]. Chebyshev type I filters were used, and the passband frequency parameter of the nth sub-band was [6 + 8 × (n-1) Hz, 92 Hz], the stopband frequency parameter was [4 + 8 × (n-1) Hz, 94 Hz], where n = 1, 2, …, 8 in this study. The data were segmented according to the time labels that set at the beginning of each trial, and SSVEP samples were generated from 140 ms to 440 ms after the time label of each channel. P300 samples were extracted from 50 ms to 800 ms after the time label.

2.5 Classification Algorithm and Decision Fusion Method

In this study, SWLDA and eTRCA were used to identify P300s and SSVEPs, respectively. SWLDA has achieved good performance for P300 recognition [13]. The algorithm combines forward and backward stepwise regression to select more important features from the original feature space, and realizes data dimensionality reduction. Then a weighted combination for the selected features will be con-ducted using the least squares regression. Lastly, the combined features will be used for class prediction. TRCA finds spatial filters to maximize the reproducibility among trials of SSVEPs [17]. The eTRCA is an advanced version of TRCA, which shows outstanding performance for SSVEP recognition. The eTRCA method always is used with the filter bank method, which decomposes SSVEP features into different sub-band components. Then a series of correlation coefficients would be calculated and fused, which is regarded as the decision value for classification. The details of eTRCA could be found in the study by Nakanishi et al. [17].

The target recognition among 120 commands in this study was based on 30 sub-spellers recognition and 4 sub-speller intra-commands recognition. In this hybrid BCI, SSVEP responses both had frequency-phase-encoded information of sub-speller and time-encoded information for the target command in the sub-speller, while P300 responses had time-encoded information corresponding to the target in the sub-speller. Therefore, SSVEPs could be used to identify 30 sub-spellers with specific frequency-phase through eTRCA. Both P300s and SSVEPs could be used to identify the time-locked target command in the sub-speller through SWLDA and eTRCA, respectively.

For target recognition among 4 commands in the sub-speller, two decision values would be calculated through SWLDA and eTRCA. Both values could be used for classification independently, while the fused value could improve the effects for command recognition. The best weights for these two values were determined according to the highest accuracy for the offline dataset, and the accuracy was obtained by traversing all possible values (from 0 to 1) in a step length of 0.05. Finally, we chose 0.25 as the fusion weight for P300s' decision value and 0.75 for SSVEPs' decision value.

2.6 System Performance Evaluation

ITR is commonly used to measure the performance of BCIs. The calculation formula of ITR is,

$$ITR = \left(\log_2 N + P\log_2 P + (1-P)\log_2\left(\frac{1-P}{N-1}\right)\right) \times \frac{60}{T} \tag{1}$$

where P represents the classification accuracy. N represents the number of commands. T is time-consuming to output a single command. In this study, the time for outputting a command in 1 to 5 rounds was 1.45 s, 2.20 s, 2.95 s, 3.70 s, and 4.45 s, respectively.

3 Results and Discussion

3.1 Subjects EEG Feature Analysis

EEG features across all subjects in offline experiments are shown in Fig. 2. The frequency information of P300s is mainly ranged from 1 Hz to 10 Hz, which is quite different from the frequency range of SSVEPs, hence we used different filters to extract the corresponding features. Figure 2(a) shows the waveform of P300s (filtered from 1 to

10 Hz) at Cz from −200 ms to 800 ms. The results revealed that target stimuli can evoke higher amplitude between 200 ms and 400 ms compared to non-target stimuli. For SSVEPs, a bandpass filter from 8 to 20 Hz was applied to attenuate the effect of P300s, Fig. 2(b)(c)(d) shows the spectral analysis results of the sub-spellers encoded by 10.5 Hz, 13.5 Hz, and 16.5 Hz at POz, respectively. It shows that the frequency of evoked SSVEPs was similar to the encoded frequency for different frequency bands. Similar spectral results could also be obtained for other sub-spellers. Therefore, the specific and satisfactory P300s and SSVEPs with wider frequency bands had been evoked in this study, which could be used for target identification.

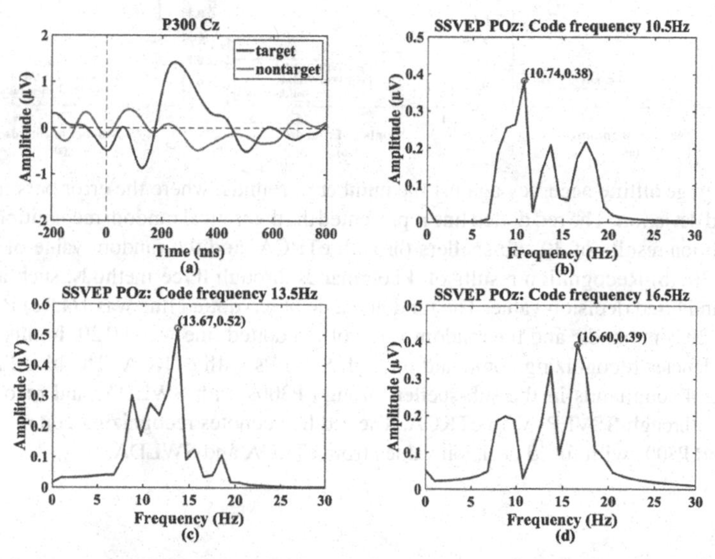

Fig. 2. EEG characterization results. The time domain or frequency domain analysis in the figure was the average results across all subjects.

3.2 Offline BCI Performance Analysis

During offline BCI performance analysis, the train set and test set were generated by leave-one-out cross-validation. Specifically, Fig. 3(a) represents the average accuracy for 30 sub-spellers, which was 64.26%, 82.78%, 90.14%, 92.91%, 94.90% for 1 to 5 rounds, respectively. Recognition results for 4 commands are shown in Fig. 3(b), the decision fusion method achieved the highest accuracy compared to eTRCA and SWLDA, and its average accuracy from 1 to 5 rounds was 93.42%, 98.05%, 99.12%, 99.29%, and 99.44%. Paired t-test analysis was used to indicate whether these differences reached a significant level in statistics. Specifically, the recognition accuracy of the decision fusion method using hybrid features was significantly higher than eTRCA using SSVEP features and SWLDA using P300 features at the first round (Hybrid vs. SSVEPs: $t_5 = 6.06, p < 0.001$, Hybrid vs. P300s: $t_5 = 8.75, p < 0.0005$). Figure 3(c) shows recognition results of 120 commands with three methods. Similarly, the decision fusion method using hybrid features was significantly better than eTRCA using SSVEP features and SWLDA using P300 features at the first round (Hybrid vs. SSVEP: $t_5 = 4.65, p < 0.005$, Hybrid vs. P300-SSVEP: $t_5 = 6.70, p < 0.001$), and its average accuracy from 1 to 5 rounds was 63.87%, 81.99%, 89.81%, 92.59%, and 94.68%, respectively.

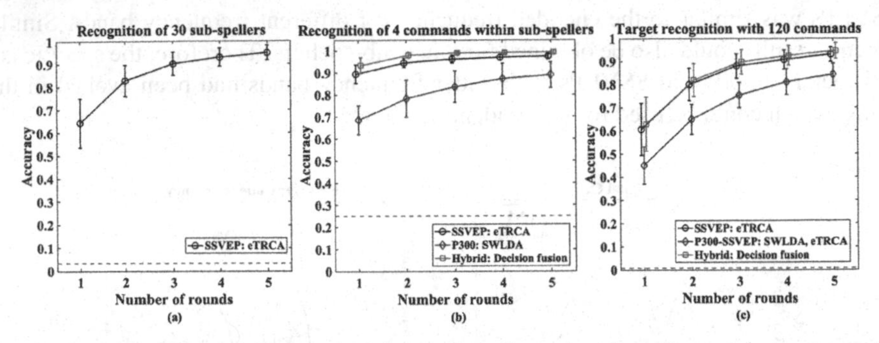

Fig. 3. Average offline accuracy against the number of rounds, where the error bars represented the standard deviation. The red dotted line represented the theoretical random recognition accuracy. (a) Recognition results of 30 sub-spellers through eTRCA, and the random value of red dotted line was 1/30. (b) Recognition results of 4 commands through three methods, such as eTRCA, SWLDA, and fused decision value. The random value of red dotted line was 1/4. (c) Recognition results of 120 commands, and the random value of red dotted line was 1/120. In this figure, the black line denotes recognizing command through SSVEPs with eTRCA. The blue line denotes recognizing 4 commands in the sub-speller through P300s with SWLDA, and recognizing 30 sub-spellers through SSVEPs with eTRCA. The red line denotes recognizing command through SSVEPs and P300s with fused decision values from eTRCA and SWLDA.

3.3 Online BCI Performance Analysis

Table 1. Online system performance.

Subject	Consuming time (s)	Selections (Correct/Total)	Accuracy (%)	ITR (bits/min)
S1	$2.20(0.7 + 2 \times 0.75)$	26/30	86.67	147.86
S2	$2.20(0.7 + 2 \times 0.75)$	27/30	90.00	156.77
S3	$1.45(0.7 + 1 \times 0.75)$	22/30	73.33	175.09
S4	$2.20(0.7 + 2 \times 0.75)$	22/30	73.33	115.40
S5	$2.20(0.7 + 2 \times 0.75)$	26/30	86.67	147.86
S6	$2.20(0.7 + 2 \times 0.75)$	28/30	93.33	166.19
Min	--	--	73.33	115.40
Max	--	--	93.33	175.09
Mean ± STD	--	--	83.89 ± 8.54	151.53 ± 20.64

The number of online experimental rounds for all subjects was determined as long as the accuracy of this round was higher than 70%, and all subjects used one or two rounds to spell one command. The command recognition method used the optimal way in offline analysis, which was the decision fusion method. Table 1 shows the online system performance across six subjects. Each subject spelled 30 commands, and the average spelling accuracy was 83.89%. Meanwhile, the maximum, minimum, and average ITR was 175.09 bits/min, 115.40 bits/min, and 151.53 bits/min, respectively. The shortest and longest time to spell out one command was separately 1.45 s and 2.2 s. It is worth noting that compared with the subjects who had experience in the BCI experiment, the inexperienced subjects S2 and S5 also had a good recognition effect. Their online spelling accuracy reached 90% and 86.67%, respectively.

4 Conclusion

This study designed a high-speed online BCI system with a large command set based on hybrid P300 and SSVEP features. A time-frequency-phase encoding strategy was used and the system encoded 120 commands in short-time with a wider SSVEP frequency band. SWLDA, eTRCA, and the decision fusion method were used to identify the target command. The decision fusion method performed the best because of ex-ploiting hybrid SSVEPs and P300s. In the results across six subjects, the online average spelling accuracy was 83.89%, the highest ITR was 175.09 bits/min, and the shortest time for spelling one command was 1.45 s. Above all, this study extended the number of commands to 120 with the wider frequency band of SSVEPs, and achieved high-speed over 100 bits/min at the same time. It provided a promising attempt to expand the number of system commands using SSVEPs and P300s, and was help to further improve the performance of BCI.

Acknowledgments. This work is supported by National Natural Science Foundation of China (No. 62106170, 62122059, 81925020, 61976152), and Introduce Innovative Teams of 2021 "New High School 20 Items" Project (2021GXRC071).

References

1. Gao, S., Wang, Y., Gao, X., Hong, B.: Visual and auditory brain- computer interfaces. IEEE Trans. Biomed. Eng. **61**(5), 1436–1447 (2014)
2. Xiao, X., Xu, M., Jin, J., Wang, Y., Jung, T.-P., Ming, D.: Discriminative canonical pattern matching for single-trial classification of ERP components. IEEE Trans. Biomed. Eng. **67**(8), 2266–2275 (2019)
3. Wang, K., Xu, M., Wang, Y., Zhang, S., Chen, L., Ming, D.: Enhance decoding of pre-movement EEG patterns for brain-computer interfaces. J. Neural Eng. **17**(1), 016033–016045 (2020)
4. Townsend, G., Platsko, V.: Pushing the p300-based brain–computer interface beyond 100 bpm: extending performance guided constraints into the temporal do-main. J. Neural Eng. **13**(2), 026024–026038 (2016)
5. Chen, X., Chen, Z., Gao, S., Gao, X.: A high-ITR SSVEP-based BCI speller. Brain-Computer Interfaces **1**(3–4), 181–191 (2014)

6. Yuan, H., He, B.: Brain-computer interfaces using sensorimotor rhythms: current state and future perspectives. IEEE Trans. Biomed. Eng. **61**(5), 1425–1435 (2014)
7. Katyal, A., Singla, R.: A novel hybrid paradigm based on steady state visually evoked potential & P300 to enhance information transfer rate. Biomed. Signal Process. Control **59**, 101884–101898 (2020)
8. Xu, M., Xiao, X., Wang, Y., Qi, H., Jung, T.-P., Ming, D.: A brain–computer interface based on miniature-event-related potentials induced by very small lateral visual stimuli. IEEE Trans. Biomed. Eng. **65**(5), 1166–1175 (2018)
9. Farwell, L.A., Donchin, E.: Talking off the top of your head: toward a mental prosthesis utilizing event-related brain potentials. Electroencephalogr. Clin. Neurophysiol. **70**(6), 510–523 (1988)
10. Jing, J., et al.: An adaptive P300-based control system. J. Neural Eng. **8**(3), 036006–036019 (2011)
11. Liu, Y., Wei, Q., Lu, Z.: A multi-target brain-computer interface based on code modulated visual evoked potentials. Public Library of Science **13**(8), 0202478–0202494 (2018)
12. Chen, Y., Yang, C., Ye, X., Ye, X., Wang, Y., Gao, X.: Implementing a calibration-free ssvep-based bci system with 160 targets. J. Neural Eng. **18**(4), 046094–046103 (2021)
13. Xu, M., Han, J., Wang, Y., Jung, T.-P., Ming, D.: Implementing over 100 command codes for a high-speed hybrid brain-computer interface using concurrent P300 and SSVEP features. IEEE Trans. Biomed. Eng. **67**(11), 3073–3082 (2020)
14. Pfurtscheller, G., Allison, B.Z., Brunner, C., Bauernfeind, G., et al.: The hybrid bci. Front. Neurosci. **4**(30), 30–40 (2010)
15. Xu, M., Qi, H., Wan, B., Yin, T., Liu, Z., Ming, D.: A hybrid BCI speller paradigm combining P300 potential and the SSVEP blocking feature. J. Neural Eng. **10**, 026001–026013 (2013)
16. Chen, X., Wang, Y., Nakanishi, M., Jung, T.-P., Gao, X.: Hybrid frequency and phase coding for a high-speed SSVEP-based BCI speller. In: 2014 36th Annual International Conference of the IEEE Engineering in Medicine and Biology Society, pp. 3993–3996 (2014)
17. Nakanishi, M., Wang, Y., Chen, X., Wang, Y.-T., Gao, X., Jung, T.-P.: Enhancing detection of SSVEPs for a high-speed brain speller using task-related component analysis. IEEE Trans. Biomed. Eng. **65**(1), 104–112 (2018)
18. Chen, X., Wang, Y., Gao, S., Jung, T.-P., Gao, X.: Filter bank canonical correlation analysis for implementing a high-speed SSVEP-based brain-computer interface. J. Neural Eng. **12**(4), 046008–046022 (2015)

Author Index

Printed in the United States
by Baker & Taylor Publisher Services

Printed in the United States
by Baker & Taylor Publisher Services